Vertebrate Cell Culture I

C. Heath, G. Belfort
Immobilization of Suspended Mammalian Cells:
Analysis of Hollow Fiber and Microcapsule Bioreactors

A. Shahar, S. Reuveny
Nerve and Muscle Cells on Microcarriers in Culture

M. Butler
Growth Limitations in Microcarrier Cultures

P. Bagnarelli, M. Clementi
Serum-Free Growth of Human Hepatoma Cells. A Review

U. Bjare
Serum-Free Cultivation of Lymphoid Cells

P. Sureau
Rabies Vaccine Production in Animal Cell Cultures

P. J. Radlett
The Use of BHK Suspension Cells for the Production of Foot and Mouth Disease Vaccines

J. B. Griffiths, A. Electricwala
Production of Tissue Plasminogen Activators from Animal Cells

AKADEMIE-VERLAG BERLIN

Vertebrate Cell Culture I

Vertebrate Cell Culture I

Managing Editor: A. Fiechter

with 39 Figures and 28 Tables

Akademie-Verlag Berlin 1987

Die Originalausgabe erscheint im Springer-Verlag Berlin Heidelberg New York
als Volume 34 der Schriftenreihe
Advances in Biochemical Engineering/Biotechnology

Vertrieb ausschließlich für die DDR und die sozialistischen Länder Akademie-Verlag Berlin

ISBN 3-540-17626-8 Springer-Verlag Berlin Heidelberg New York Tokyo
ISBN 0-387-17626-8 Springer-Verlag New York Heidelberg Berlin Tokyo

ISBN 3-05-500505-8

Erschienen im Akademie-Verlag Berlin, DDR-1086 Berlin, Leipziger Straße 3—4
Lizenznummer: 202 · 100/522/87
Printed in the German Democratic Republic
Gesamtherstellung: VEB Druckerei „Thomas Müntzer", 5820 Bad Langensalza
LSV 1345
Bestellnummer: 763 757 8 (3070/34)
11800

Table of Contents

Immobilization of Suspended Mammalian Cells: Analysis of Hollow Fiber and Microcapsule Bioreactors

Carole Heath and Georges Belfort
Department of Chemical Engineering and Environmental Engineering,
Rensselaer Polytechnic Institute, Troy, New York 12180-3590, U.S.A.

Model equations of substrate mass transfer and uptake have been formulated for two bioreactor systems: microcapsules and a hollow fiber reactor. Assumptions include time independence, Fickian diffusion, homogenous cell suspension, and kinetics described by the zero and first order limits of the Monod equation. Glucose and oxygen were the substrates chosen for investigation of the kinetic and diffusion limitations. For microcapsules, the resulting radial concentration profiles indicated the possibility of a necrotic core due to insufficient substrate in those cases where diffusion is low and/or uptake is high. The model equations provide a means of estimating the maximum capsule radius which will allow adequate diffusion of nutrients to all of the contained cells. The simulated concentration gradients developed for a hollow fiber reactor demonstrate that diffusion limitations may exist at the far end of the reactor. Concentration gradients are shown both in the fiber lumens (in the axial direction) as well as in the shell space (in the radial direction). Variation of model parameters provides information on system specifications to avoid these diffusion limitations.

1 Introduction

The development of novel bioreactors has flourished in recent years as efforts to produce more efficient and economical systems continue. Bioreactors are used for production of many different compounds from plant [28], microbial and animal cells, as well as from isolated enzyme systems. Conventionally, bioreactors have consisted of batch and continuous flow reactors, typically known as stirred tanks and chemostats, respectively. To improve the performance of these reactors, designs have focused on increasing productivity per unit volume and reducing the amount of expensive downstream processing. Immobilization of cells with various types of barriers or supports can satisfy both criteria. For suspended cells, immobilization primarily consists of two options [19]:

a. entrapment of cells within a porous matrix; and
b. containment of cells behind a barrier.

The first option includes porous supports such as ceramics and gels, which can be either preformed or formed around the cells [26]. This option will not be discussed further. The second option includes processes which retain the cells behind a barrier, typically a membrane. Hollow fiber and flat sheet reactors and microcapsules all utilize membranes for this purpose. In hollow fiber and flat sheet reactors, the membrane is preformed; in microcapsules, the membrane is formed around the cells. Because each of these systems protects the delicate outer membranes of animal cells from the shear forces which result from media flow or stirring, they have been used for cultivation of mammalian cells [3,9,13,14,17,19,23,24,27]. Although most animal cells are anchorage dependent for growth, transformed cells, such as hybridoma cells, are grown in suspension culture. This is possible with the hollow fiber and flat sheet reactors as well as with the microcapsules. Each method, however, provides a different environment for the suspended cells, an environment which ultimately determines the productivity of the system.

Of major concern to the optimization of conversions in these bioreactors is whether, by instituting the separation or immobilization of the cells, the reactions are governed by kinetic or diffusional control. Models to determine the answer to this question can also provide insight toward shifting the mechanisms of control when the rate of production is dictated by diffusional limitations. This paper addresses current analyses of the problem and provides results from solutions of simplified continuity equations for the hollow fiber bioreactor and for the microcapsule using two substrates, glucose and oxygen.

2 Immobilization

2.1 Microcapsules

The encapsulation of *living* cells was first accomplished by Lim and Moss [25]. Jarvis and Grdina [17] have developed a modification of the original procedure which is patented under the trade name Encapcel. The process involves formation of a spherical polyanionic gel mold containing cells, onto which is deposited a polymeric membrane. The inner gel can then be liquified and allowed to diffuse out of the capsule leaving behind the membrane and the contained cells. The membrane, typically calcium or sodium alginate and poly-l-lysine, can be degraded, allowing release of the cells as desired. Both the porosity of the membrane and the size of the microcapsule can be varied to accommodate many reactant-product systems. Capsule diameters from 20 nm to 2 mm are possible. The porosity can be varied by several orders of magnitude, from retention of glucose (180 Da) to permeation of IgG (155,000 Da).

2.2 Hollow Fiber Reactors

The successful use of hollow fiber reactors for the cultivation of mammalian cells was first reported by Knazek et al. in 1972 [22]. Since then, several others have published results for mammalian cells using hollow fiber systems (Knazek et al., 1974; Calabresi et al., 1981; Ku et al., 1981; Wiemann et al., 1983; Hopkinson, 1985; Tharakan and Chau, 1986; Altshuler et al., 1986). Many different types of mammalian cells have been cultured in hollow fiber reactors (Table 1) and various modules have been used for the experiments (Table 2). The core of the reactor consists of many hollow fiber membranes, the composition and porosity of which can be varied for different systems. The culture media usually flows through the lumens of these fibers, allowing diffusion of nutrients from the media to the extracapillary, or shell, space which houses the cells. Depending on membrane porosity, certain cell byproducts may also diffuse from the shell space into the fiber lumens where they are carried away by the media. Hollow fiber systems vary in complexity and function, incorporating recirculation loops with various bleeds, pumps, supplies, gauges, controllers, etc. The hollow fiber cartridge dimensions as well as the characteristics of the membrane can be varied to provide versatility for reacting systems. A variant of the cylindrical shell design,

Table 1. Mammalian cell types cultured successfully in hollow fiber bioreactors[a]

Organism	Tissue, cell type		
Chick	Embryo, Primary Culture	Human	Foreskin, HFF/HR 218
Duck	Embryo, Primary Culture		Pituitary Tumor, Primary Culture
Mouse	Connective Tissue, L-929		Choriocarcinoma, JEG-7
	Embryo, 3T3/SV3T3		Breast Carcinoma, BT-20/HBT-3
	Myeloma/Spleen Hybridoma		B Lymphoma, NC-37
	Mammary Tumor, 66/67/168/G8H		Colon Adenocarcinoma, LS-174T
Rat	Pituitary Tumor, MtTW5		Colorectal Adenocarcinomas, SW403/480/620/707
	Liver Hepatoma, Reuber H4-11-E		Colon Carcinoma, DLD-1
Hamster	Kidney, BHK-21		Lymphoma, Myeloma, Leukemia, Melanoma, and Colon Carcinoma
	Ovary, Ha-1		Cervical Carcinoma, HeLa
Monkey	Lung, Rhesus Diploid		Lung, WI-38
	Kidney, Vero		Myeloma/Spleen Hybridoma
	Kidney, Rhesus MK-2		Liver Hepatoma, Alexander PLC/PRF/5

[a] After Hopkinson [14]

Table 2. Growth of mammalian cells in hollow fiber modules

Species	Hollow fiber module						
	MWCO $\times 10^{-3}$	Area cm^2	Material	Volume cm^3	Length of experiment	Ref. Authors	Year
Monkey fibroblasts rat pituitary tumor Human choriocarcinoma Human breast carcinoma	30, 50	46–100	CA, PVC-acrylic, silicone-polycarbonate	1.5–2.5	Several weeks	22, 23, 23a)	1972, 74,79
Reuber hepatoma (rat)	30, 50	75	CA, PVC-acrylic	1.5–2.5	20–44 h	42)	1975
Human adeno-carcinoma of the colon, human melanoma	10	—	Polysulfone	2.5	—	7a, 28b)	1977,78
Moloney murine leukemia virus	100	60	Polysulfone	2.5	11 d	28a)	1978
Human colon adenocarcinoma	10–50	—	—	2.5	30–40 d	31a)	1979
Vero rhesus monkey kidney	50	—	Polysulfone, polyacrylo-nitrile	2.5	70 h	24)	1981
Mouse hybridoma	10	100	Polysulphone	2.5	23 d	4)	1986
	50	100	Polysulphone	2.5	23 d		
	100	100	Polysulphone	2.5	23 d		
Hamster fibroblasts	100	1150	Polysulphone	145–46	2 weeks	34)	1986

the flat-bed hollow fiber reactor was proposed by Ku et al. [24] in order to overcome diffusion limitations inherent in cartridge-type reactors (see further discussion below). Another variant is the radial flow hollow fiber bioreactor reported by Tharakan and Chau [34].

3 Theory

3.1 General

Several papers describing theoretical models for concentration profiles in hollow fiber reactors have been published. A summary of a few significant analyses of bioreactor dynamics is presented in Table 3. All of these studies except that of Schonberg and Belfort [32] have assumed that radial convection through the understructure of the membrane is not important. Later it will be seen that this assumption may not be justified. Waterland et al. [36] provided a theoretical model and supporting experimental data [37] for a first order enzymatic reaction at steady state using asymmetric hollow fiber membranes. The experimental results correlated well with model predictions. A similar approach was taken by Kim and Cooney [20] They considered the general

Table 3. Modeling of hollow fiber bioreactor dynamics

System	Kinetics[a]	Comments	Ref.
Enzymes	0, 1	Enzymes in lumen, well-mixed shell	31)
Enzymes	mm	Coated inner lumen	15)
Enzymes	mm	Enzymes in spongy matrix, arbitrary axial concentration in lumen as a function of z	36)
Enzymes	mm	Enzymes in lumen, well-mixed shell	12)
Enzymes	1	Enzymes in spongy matrix, axial concentration gradient in lumen	20)
Whole cells	0, 1, mm	Cells in shell space, well-mixed lumen	38–40)
Whole cells	any	Cells in shell space	8)
Whole cells, enzymes	0, 1	Catalyst in shell space, well-mixed lumen, impose radial convection	32)

[a] 0, 1, mm, and any designate zero order, first order, Michaelis-Menten, and any reaction rate laws, respectively

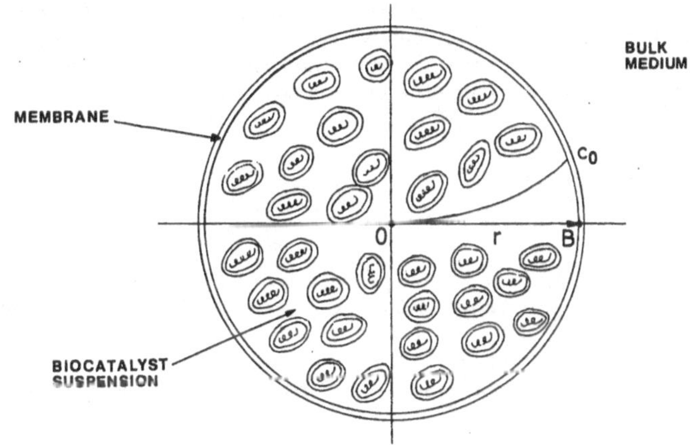

a

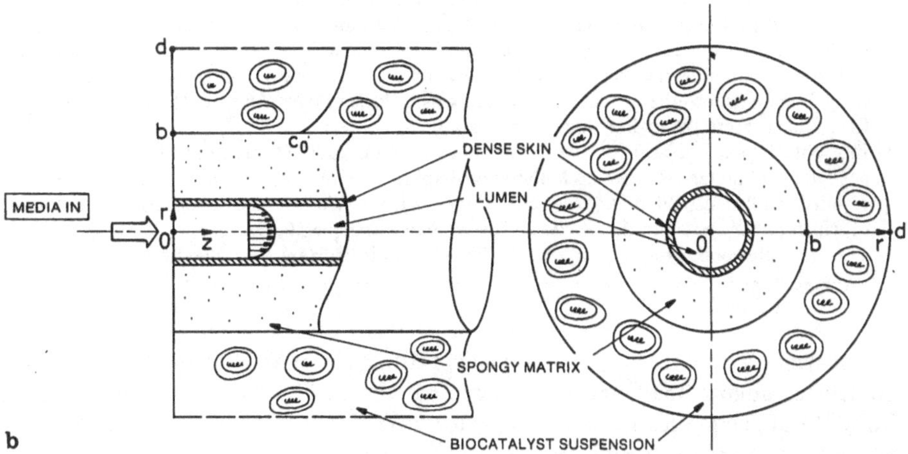

b

Fig. 1 a and b. Typical expected substrate concentration profiles based on model assumptions for (**a**) microcapsules and (**b**) a hollow fiber reactor

Table 4. Assumptions used in analysis

General Assumptions (apply to both hollow fibers and microcapsules)

1. The suspension is homogeneous.
2. The system is isothermal.
3. Diffusion of the substrate within the cell suspension can be described by Fick's Law using an effective diffusion coefficient, D_e.
4. The membrane provides negligible resistance to diffusion compared to the resistance to mass transfer in the cell suspension. This assumption has been justified by previous studies [18].
5. The system is at steady state with respect to substrate conversion. This assumption is justified by Webster and Shuler's [40] investigation into transient analysis of concentration profiles. They found that steady state profiles of substrate concentration are established rapidly (in seconds), subject to variation of system parameters.
6. The kinetics of substrate utilization can be described by limiting forms of the Michaelis-Menten (or Monod) equation:

$$R_a = \frac{V_m C_a}{K_m + C_a}$$

where V_m is the maximum rate of reaction and K_m is the Michaelis constant. The first and zero order limits are typically used to describe the kinetics of whole cell reactions [19]. Despite the fact that Kleinstreuer and Poweigha [21] maintain, without justification, that Michaelis-Menten kinetics are not applicable to whole cells, the assumption will be used here to provide a simple approximation. For the zero order limit, the expression becomes

$$R_a = V_m, C_a \gg K_m$$

The first order limit expression is

$$R_a = V_m C_a / K_m, C_a \gg K_m$$

Microcapsule Assumptions

7. There is no convective fluid motion inside the particles. The validity of this assumption depends on the rate of stirring of the bath solution and on the permeability and flexibility of the microcapsules.
8. Diffusion within the spheres is a geometrical function of radius only.
9. The bulk solution (culture media) is well-mixed and is maintained at a constant substrate concentration.

Hollow fiber assumptions

7. Convective flow occurs in the fiber lumens only, and is represented by an average axial velocity, V_z. Other velocity components are negligible. This last sentence may not be justified when large pressure drops occur along the fiber length. Waterland et al. [37] found that axial enzyme transport occurred in the shell space of hollow fiber reactors due to radial fluid flow across the early fiber walls into the extracapillary space which flowed down the module and reentered the fiber lumen where the hydrostatic pressure gradient was less.
8. In the shell space, diffusion occurs in the radial direction only. Diffusion in the fiber lumens is negligible. Although this assumption greatly simplifies the problem solution, it may not be justified in all cases. When large concentration gradients exist in the axial direction, examples of which are presented in the results, diffusion may not be negligible.
9. The substrate concentration profile in the fiber lumen is constant, i.e., plug flow.

case in which axial concentration gradients exist. Webster and Shuler [38] have derived models for hollow fiber enzyme reactors and whole cell hollow fiber reactors (with Rony [39]) utilizing effectiveness factors, for both first and zero order kinetics. The effectiveness factor is a ratio of the actual reaction rate to the rate without mass transfer limitations. Webster and Shuler [40] have also developed a model for simulating transient substrate concentration profiles for whole cell hollow fiber reactors. They considered a less general case in which axial gradients do not exist. For systems with short residence times and frequent recycle, the overall system resembles a continuously stirred tank reactor.

Van Heuven et al. [35] published solutions to the continuity equation for enzymes in a spherical carrier material for both zero and first order rate limits.

Most of the papers mentioned above simulate concentration profiles for the reactors as a function of the Thiele modulus or a modified Thiele modulus. The Thiele modulus is a function of radius, diffusivity, and the kinetic constants, and relates kinetics to internal diffusion resistance. Because the Thiele modulus is a nondimensional number, correlation expressed as a function of the modulus can be applied to many systems. The approach taken in this paper is slightly different. Instead of presenting a general correlation for all systems, this paper will demonstrate development of the model equations and apply them to particular systems. For the hollow fiber bioreactor, the characteristics and dimensions of a typical commercial module will be used; for the microspheres, three different radii will be investigated.

As with the previously published models, the mathematical development will begin with the continuity equations and a list of assumptions summarized in Table 4.

The development of models for substrate concentration profiles as functions of radius and fiber length (hollow fibers only) follows. Each system is solved for zero and first order limits of Michaelis-Menten kinetics. Figure 1 illustrates the coordinate systems and demonstrates expected substrate concentration profiles in each type of bioreactor based on the listed assumptions.

3.2 Microcapsules

From Bird et al. [6], the equation of continuity of any component in spherical coordinates with constant density and diffusivity is

$$\frac{\partial C}{\partial t} + V_r \frac{\partial C}{\partial r} + V_\theta \frac{1}{r} \frac{\partial C}{\partial \theta} + V_\varphi \frac{1}{r \sin\theta} \frac{\partial C}{\partial \varphi}$$

$$= D_e \left(\frac{1}{r^2} \frac{\partial}{\partial r} \left(r^2 \frac{\partial C}{\partial r} \right) + \frac{1}{r^2 \sin\theta} \frac{\partial}{\partial \theta} \left(\sin\theta \frac{\partial C}{\partial \theta} \right) + \frac{1}{r^2 \sin^2\theta} \frac{\partial^2 C}{\partial \varphi^2} \right) - R \quad (1)$$

Note that the effective diffusivity has been substituted for the binary diffusivity. R represents the uptake term. With the given assumptions, the equation reduces to

$$D_e \left(\frac{1}{r^2} \frac{d}{dr} \left(r^2 \frac{dC}{dr} \right) \right) - R = 0 \quad (2)$$

This can be expanded to

$$D_e \left(\frac{d^2C}{dr^2} + \frac{2}{r} \frac{dC}{dr} \right) - R = 0 \quad (3)$$

For the zero order kinetic limit this becomes

$$\frac{d^2C}{dr^2} + \frac{2}{r}\frac{dC}{dr} - \frac{V_m}{D_e} = 0 \tag{4}$$

The boundary conditions are

$$\text{at} \quad r = R\,, \qquad C(R) = C_0 \tag{5a}$$

$$\text{at} \quad r = 0\,, \; \frac{dC(r)}{dr} = 0 \tag{5b}$$

The solution is easily obtained [35] and is given as

$$C(r) = C_0 + \frac{V_m}{6D_e}(r^2 - R^2) \tag{6}$$

For the first order kinetic limit, Eq. (3) becomes

$$\frac{d^2C}{dr^2} + \frac{2}{R}\frac{dC}{dr} - \frac{V_m}{D_e K_m} C = 0 \tag{7}$$

The boundary conditions are the same as given in Eqs. (5a) and (5b). The solution to Eq. (7) is not as easily solved but is given elsewhere [6, 35] as

$$C(r) = C_0 \left(\frac{R}{r}\right) \frac{\sinh\left(r\sqrt{V_m/K_m D_e}\right)}{\sinh\left(R\sqrt{V_m/K_m D_e}\right)} \tag{8}$$

The solution is undefined at $r = 0$, however the limit as $r \to 0$ exists.

The diffusion and uptake of a substrate at the zero and first order kinetic limits of the Monod equation is described by Eqs. (6) and (8), respectively.

3.3 Hollow Fiber Reactor

The equation of continuity of a component in cylindrical coordinates with constant density and diffusivity applies to hollow fiber systems and is given by

$$\frac{\partial C}{\partial t} + \left(V_r \frac{\partial C}{\partial r} + V_\theta \frac{1}{r}\frac{\partial C}{\partial \theta} + V_z \frac{\partial C}{\partial z}\right) = D_e \left(\frac{1}{r}\frac{\partial}{\partial r}\left(r \frac{\partial C}{\partial r}\right) + \frac{1}{r^2}\frac{\partial^2 C}{\partial \theta^2} + \frac{\partial^2 C}{\partial z^2}\right) - R \tag{9}$$

With the assumptions given in Table 4 this reduces to

$$V_z \frac{\partial C}{\partial z} = D_e \frac{1}{r}\frac{\partial}{\partial r}\left(r \frac{\partial C}{\partial r}\right) - R \tag{10}$$

This can be simplified to

$$V_z \frac{\partial C}{\partial z} = D_e \left(\frac{\partial^2 C}{\partial r^2} + \frac{1}{r} \frac{\partial C}{\partial r} \right) - R \tag{11}$$

In this model, the hollow fiber bioreactor consists of two regions, each with different transport mechanisms. Region 1 includes the lumen and the fiber membrane. The concentration profile is assumed to be constant across a fiber cross section in this region. This includes the dense region and spongy support structure for asymmetric fibers. Region 2 consists of the shell space or annulus around the fiber in which the suspension of catalyst (cells or enzymes) is located. Diffusion is assumed to be the only significant means of transport in the outer region.

Region 1: Lumen and membrane $0 \leqq r \leqq A$
Region 2: Annular shell space $A \leqq r \leqq B$

The simplest approach is to first solve Eq. (11) for Region 2 at any given axial distance along the reactor. In this region $V_z = 0$ and Eq. (11) becomes

$$\frac{d^2C}{dr^2} + \frac{1}{r} \frac{dC}{dr} - \frac{R}{D_e} = 0 \tag{12}$$

For the zero and first order kinetic limits this becomes

$$\frac{d^2C}{dr^2} + \frac{1}{r} \frac{dC}{dr} - \frac{V_m}{D_e} = 0 \tag{13}$$

and

$$\frac{d^2C}{dr^2} + \frac{1}{r} \frac{dC}{dr} - \frac{V_m C}{K_m D_e} = 0\,, \tag{14}$$

respectively. The boundary conditions are

$$\text{at} \quad r = A\,, \qquad C(A) = C(z) \tag{15a}$$

$$\text{at} \quad r = B\,, \qquad \frac{dC}{dr} = 0 \tag{15b}$$

Equation (15a) indicates that at the membrane surface the concentration is a function of z only. Equation (13) can be solved to yield an expression of concentration as a function of r for the zero order reaction limit at any given z.

$$C(r) = 1/2(r^2/2 - B^2 \ln(r) + B^2 \ln(A) - A^2/2)\, V_m/D_e + C(z) \tag{16}$$

With this expression, the concentration in Region 1 as a function of axial distance can be determined. Consider a flux balance over a small length Δz of a membrane fiber. The convective flux into the region must equal the sum of the convective and diffusive fluxes out of the region at any given instant.

$$N_z|_z \pi A^2 - N_z|_{z+\Delta z} \pi A^2 - N_r\Big|_{\substack{r=A \\ z=z*}} 2\pi A \Delta z = 0$$

z* is the distance midway between z and z + Δz.
Rearranging and taking the limit as $\Delta z \to 0$ yields

$$\frac{d}{dz}(N_z) = \frac{-2}{A} N_r\Big|_{r=A} \tag{17}$$

Substituting the expressions $N_z = V_z C$ and $N_r = -D_e \frac{\partial C}{\partial r}$ into Eq. (17) results in

$$\frac{\partial C}{\partial z} = \frac{2nD_e}{AV_z} \frac{\partial C}{\partial r}\Big|_{r=A} \tag{18}$$

where n accounts for the number of fibers in the reactor. Equation (16) can be differentiated with respect to r and evaluated at r = A. This is substituted into Eq. (18) which can then be solved for C(z) using the following boundary condition

$$C(z = 0) = C_0 \tag{19}$$

The axial concentration profile as a function of axial distance z is given by

$$C(z) = abz + C_0 \tag{20}$$

where

$$a = nV_m/AV_z \tag{21}$$

and

$$b = (A - B^2/A) \tag{22}$$

Equations (16) and (20) provide the radial and axial solutions, respectively, for the zero order limit.

The solution of the first order limit in Eq. (14) requires the use of modified Bessel functions of the first and second kind (see Appendix for details). The radial concentration as a function of r at a given axial length z is described by

$$C(r) = C(z)\,(K_1(Bx)\,I_0(rx) + I_1(Bx)\,K_0(rx))/(I_0(Ax)\,K_1(Bx) + I_1(Bx)\,K_0(Ax)) \tag{23}$$

where I and K are modified Bessel functions of the first and second kinds, respectively. The number subscript refers to the function order. The parameter x is given as

$$x = (V_m/D_e K_m)^{1/2} \tag{24}$$

Taking the derivative of Eq. (23), evaluating at r = A, and substituting it into Eq. (18), similar to the procedure for deriving the zero order limit equation, yields an expression for axial concentration as a function of z.

$$C(z) = C_0 \exp [dz(K_1(Bx)\ I_1(Ax) - I_1(Bx)\ K_1(Ax))/(I_0(Ax)\ K_1(Bx) + I_1(Bx)\ K_0(Ax))] \tag{25}$$

where

$$d = 2nD_e/V_z A \tag{26}$$

Equations (23) and (25) provide radial and axial concentration profiles, respectively, for the first order kinetic limit.

Table 5. Parameter values used to determine concentration profiles

Parameter	Source
Glucose	
$C_o = 1.11 \times 10^{-5}$ moles ml^{-1}	[33]
$D_e = 5 \times 10^{-7}$ cm^2 s^{-1}	[11]
$D_e = 1 \times 10^{-6}$ cm^2 s^{-1}	[16]
Oxygen	
$C_o = 1 \times 10^{-6}$ moles ml^{-1}	[5]
$C_o = 1 \times 10^{-7}$ moles ml^{-1}	[10]
$D_e = 2 \times 10^{-5}$ cm^2 s^{-1}	[13]
$D_e = 2 \times 10^{-6}$ cm^2 s^{-1}	[5]
Hollow Fibers	
$V_z = 0.167$ cm s^{-1} (axial velocity)	[33] values for the entire module. (Amicon Corp.,
L = 5.7 cm (fiber length)	Danvers, MA)
A = 0.0130 cm (fiber radius)	Simulations were run for an individual fiber
B = 0.0408 cm (annulus radius)	assuming symmetry of fiber spacing
R = 0.5 cm (module radius)	
n = 150 (number of fibers)	
Microcapsules	
$B = 5 \times 10^{-3}$ cm (capsule radius)	Values chosen to range around reported values [17]
$B = 1.5 \times 10^{-2}$ cm (capsule radius)	
$B = 2.5 \times 10^{-2}$ cm (capsule radius)	
Kinetics	
$V_m = 1 \times 10^{-7}$ moles ml^{-1} s^{-1} [a]	Chosen to represent high rates of uptake
$V_m = 1 \times 10^{-8}$ moles ml^{-1} s^{-1} [a]	
$V_m = 1 \times 10^{-9}$ moles ml^{-1} s^{-1}	
$V_m = 1 \times 10^{-10}$ moles ml^{-1} s^{-1}	[3,13,30]
$V_k = 0.01$ s^{-1}	[8]
$V_k = 0.10$ s^{-1}	[40]
$V_k = 1.00$ s^{-1}	[40]

[a] These values were used with every microcapsule system because smaller values resulted in flat concentration profiles within the capsule, i.e., kinetic control

4 Computations

Four computer programs were written to solve the equations derived to describe concentration as functions of axial and radial distance. The four programs represent the following four cases:

a. microcapsules — zero order kinetic limit (Eq. (6))
b. microcapsules —first order kinetic limit (Eq. (8))
c. hollow fibers — zero order kinetic limit (Eqs. (16) and (20))
d. hollow fibers — first order kinetic limit (Eqs. (23) and (25))

Concentration profiles for the microcapsules, for oxygen and glucose, were determined using three capsule radii, chosen to constitute a size range around a reported radius [17]. As a typical hollow fiber module, dimensions were taken from

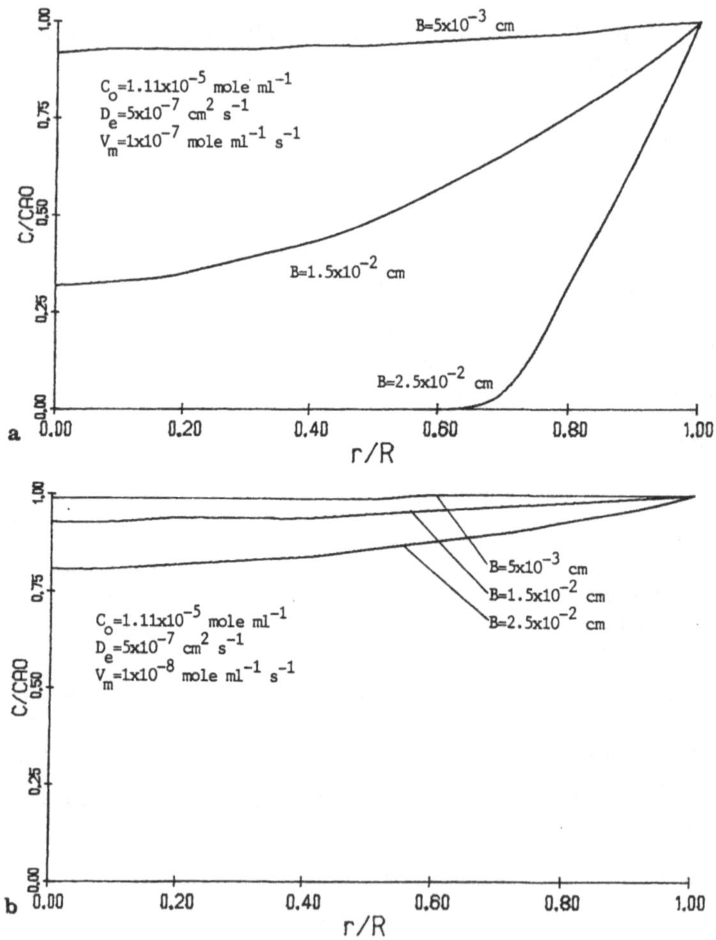

Fig. 2a and b. Radial concentration profiles for glucose in microcapsules assuming a zero order kinetic limit for substrate consumption. The maximum reaction rate is (**a**) 1×10^{-7} mole ml^{-1} s^{-1}; (**b**) 1×10^{-8} mole ml^{-1} s^{-1}

commercial literature (Amicon Corp., MA). These values are listed in Table 5 and apply to every hollow fiber result unless otherwise indicated. For the hollow fibers, the axial concentration profiles were determined first. The radial concentration gradients were then determined for the entrance, mid-point and exit of the module. Typical values of the adjustable parameters, taken both from the literature and from results obtained in our laboratory as indicated in Table 5, were varied to investigate the effects of changing the magnitudes of substrate transport and uptake. These parameters include bulk substrate concentration, substrate diffusivity, reaction rates and capsule radius, as appropriate.

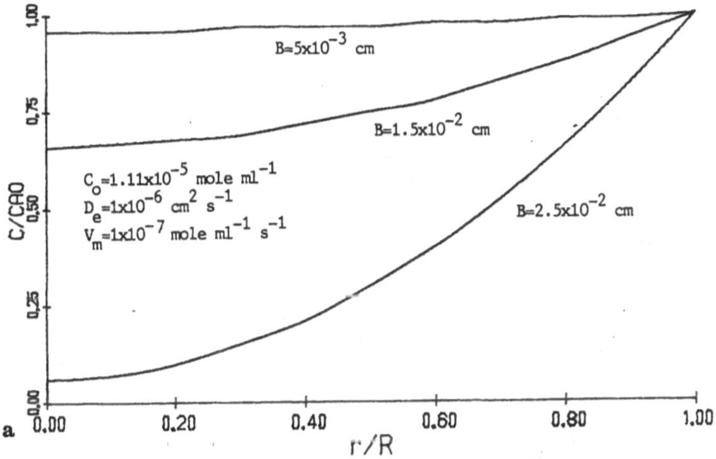

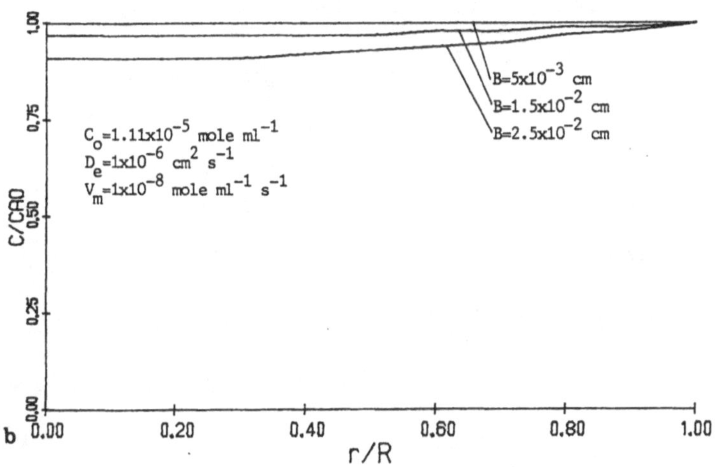

Fig. 3a and b. Radial concentration profiles for glucose in microcapsules assuming a zero order kinetic limit for substrate consumption; same as Fig. 2, except that the effective diffusivity has been increased. The maximum reaction rate is (**a**) 1×10^{-7} mole ml^{-1} s^{-1}; (**b**) 1×10^{-8} mole ml^{-1} s^{-1}

5 Results

5.1 Microcapsules

Figures 2 and 3 depict radial concentration profiles for glucose in microcapsules assuming a zero order limit to the Michaelis-Menten rate law. Figure 2 demonstrates the drastic effects of different values of the maximum reaction rate for three different capsule radii. In part a of Fig. 2 ($V_m = 1 \times 10^{-7}$ moles ml^{-1} s^{-1}) it is evident that only the two smaller spheres can provide adequate glucose for cell growth, while in Fig. 2b ($V_m = 1 \times 10^{-8}$ moles ml^{-1} s^{-1}) all three sizes are more than adequate. In Fig. 3 it is shown that increasing the effective substrate diffusivity by less than an order of magnitude has little effect on concentration

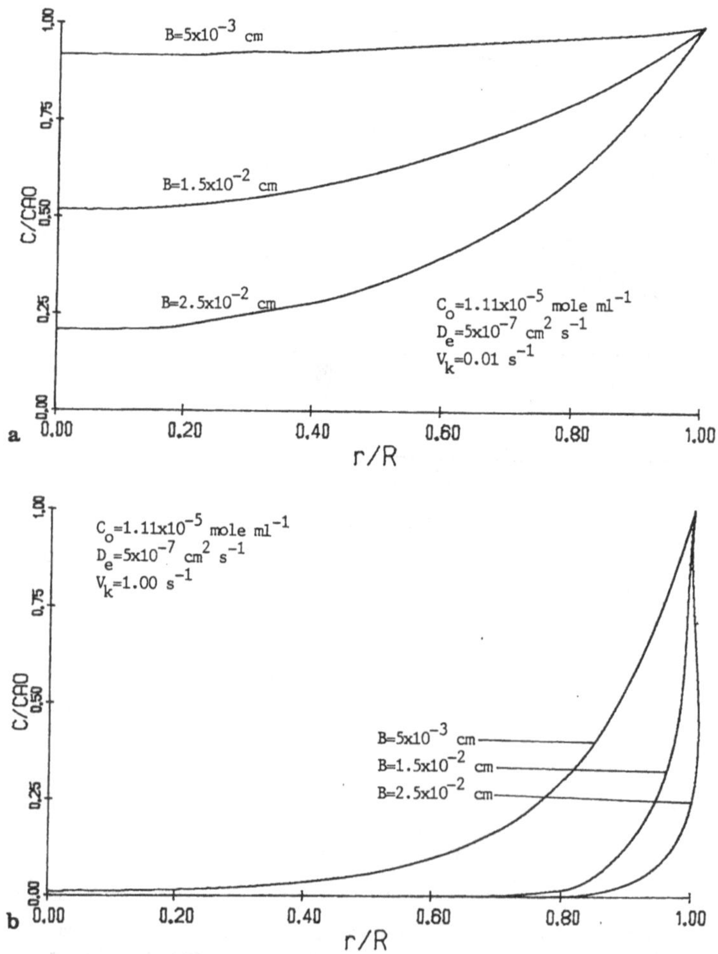

Fig. 4a and b. Radial concentration profiles for glucose in microcapsules assuming a first order kinetic limit for substrate consumption. The reaction rate is **(a)** 0.01 s^{-1}; **(b)** 1.00 s^{-1} ($V_k = V_m/K_m$)

profiles subject to the lower reaction rate (Figs. 2b and 3b) yet has a significant effect on the concentration profiles for the larger reaction rate (Figs. 2a and 3a), making all three sizes acceptable.

Figure 4 consists of radial concentration profiles for glucose assuming a first order limit to the Michaelis-Menten rate law. Again, changes in the rate constant ($V_k = V_m/K_m$) result in significant changes in the concentration profiles. The profiles in Fig. 4a show that all three sphere radii would be compatible with cell growth for a V_k of 0.01 s^{-1}. Increasing V_k to 1.00 s^{-1} is shown in Fig. 4b to result in unfavourable growth conditions for all three capsule sizes.

Radial oxygen concentration profiles for the zero order kinetic limit are shown in Figs. 5–7. Figures 5 and 6 were determined using identical parameter values except for the effective oxygen diffusivity. As was shown for glucose, increasing

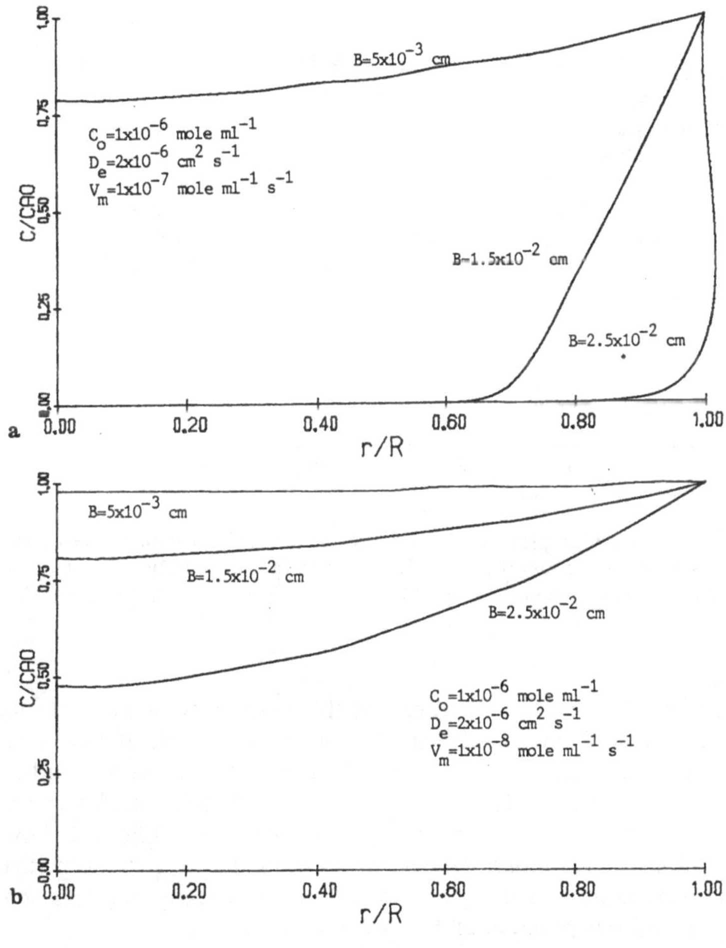

Fig. 5a and b. Radial concentration profiles for oxygen in microcapsules assuming a zero order kinetic limit for substrate consumption. The maximum reaction rate is (**a**) 1×10^{-7} mole ml^{-1} s^{-1}; (**b**) 1×10^{-8} mole ml^{-1} s^{-1}

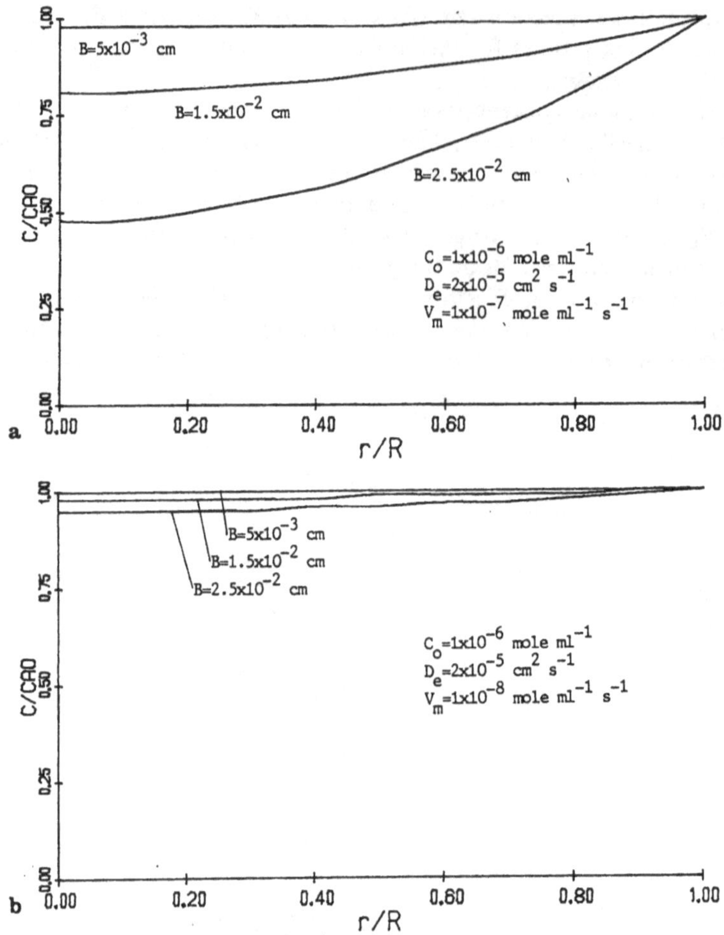

Fig. 6a and b. Radial concentration profiles for oxygen in microcapsules assuming a zero order kinetic limit for substrate consumption; same as Fig. 5, except that the effective diffusivity has been increased. The maximum reaction rate is (**a**) 1×10^{-7} mole ml^{-1} s^{-1}; (**b**) 1×10^{-8} mole ml^{-1} s^{-1}

the effective diffusivity has a dramatic effect on the availability of substrate for cellular consumption. In Fig. 5a, only the smallest sphere is capable of supporting growth throughout the capsule, yet in Fig. 6a all three sphere sizes will suffice. On the other hand, if the maximum uptake rate is 1×10^{-8} moles ml^{-1} s^{-1}, any of the three capsule radii will yield favourable growth conditions. Figure 7 shows concentration profiles for conditions identical to those used in Fig. 5 except that the bulk concentration is decreased by a factor of 10. Insufficient substrate concentration within the microcapsules in all but one case is the result.

Concentration profiles of oxygen as a function of radius in microcapsules assuming a first order limit to the rate equation are presented in Figs. 8 and 9. As can be seen in the figures, only the case where $D_e = 2 \times 10^{-6}$ cm^2 s^{-1} and

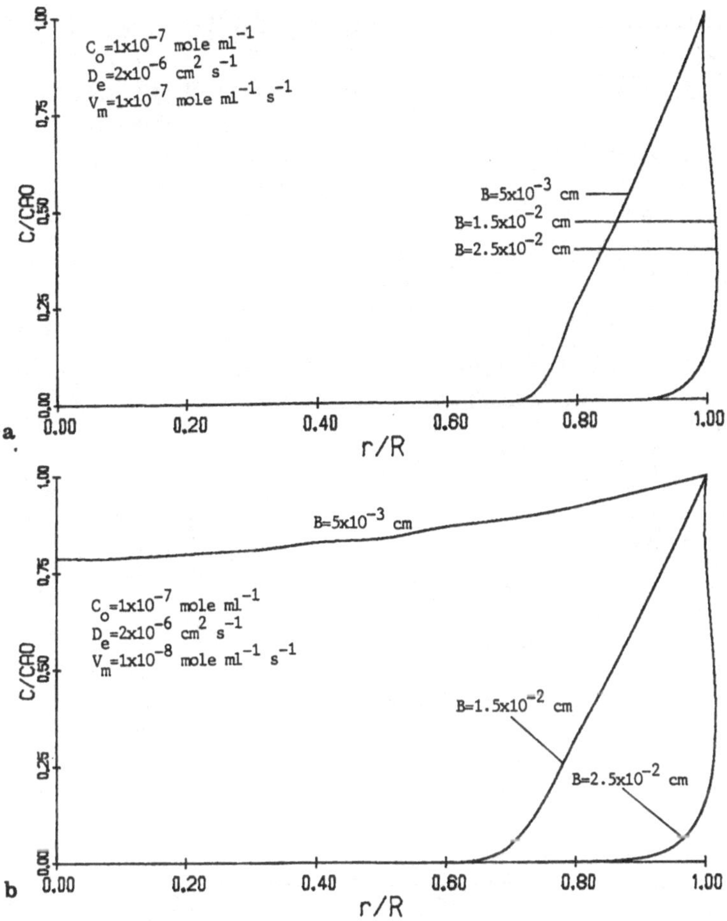

Fig. 7a and b. Radial concentration profiles for oxygen in microcapsules assuming a zero order kinetic limit for substrate consumption; same as Fig. 5, except that the bulk oxygen concentration is decreased. The maximum reaction rate is (a) 1×10^{-7} mole ml^{-1} s^{-1}; (b) 1×10^{-8} mole ml^{-1} s^{-1}

$V_k = 1.00\ s^{-1}$, i.e., low effective diffusivity and high uptake rate, do any of the sphere sizes become too large to provide enough oxygen for consumption.

5.2 Hollow Fiber Reactors

Axial and radial concentration profiles of glucose and oxygen assuming a zero order limit to the rate equation are presented in Fig. 10 for hollow fiber dimensions equal to those listed in Table 5. Figure 10a shows axial concentration profiles for the two substrates.

Using a reaction rate and substrate effective diffusivities and concentrations from Table 5, the resulting curves indicate a rapid depletion of oxygen with

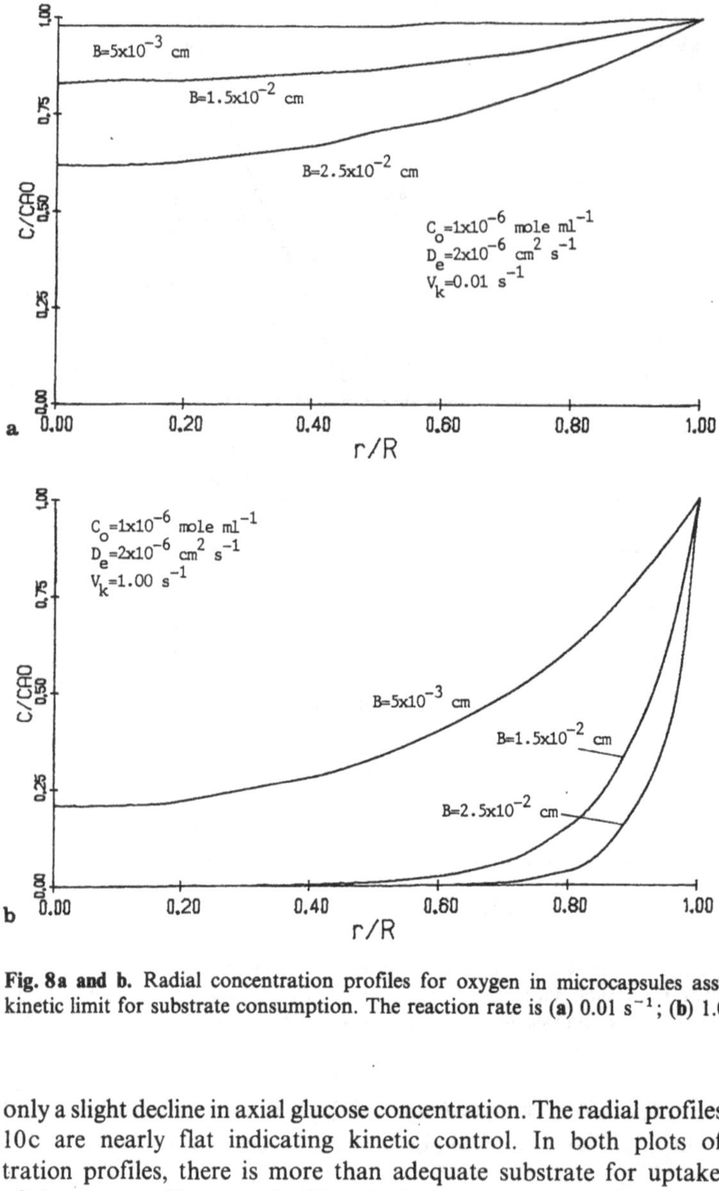

Fig. 8a and b. Radial concentration profiles for oxygen in microcapsules assuming a first order kinetic limit for substrate consumption. The reaction rate is (**a**) 0.01 s^{-1}; (**b**) 1.00 s^{-1}

only a slight decline in axial glucose concentration. The radial profiles in Figs. 10b and 10c are nearly flat indicating kinetic control. In both plots of radial concentration profiles, there is more than adequate substrate for uptake at the entrance of the reactor. However, in Fig. 10c it is apparent that there is insufficient oxygen for cells beyond the reactor entrance. The oxygen has rapidly diffused into the extracapillary space upon entering the reactor, saturating the inlet volume. As can be seen in Fig. 10a, at one-fifth of the distance along the membrane, there is negligible oxygen remaining in the fiber lumen, resulting in a large axial oxygen concentration gradient in the shell space. As will be indicated in the discussion, this would most likely result in axial diffusion of the substrate, defying hollow fiber assumption #8 in Table 4. If an axial diffusion term is included in the model, it could be determined whether there is sufficient oxygen for the entire reactor.

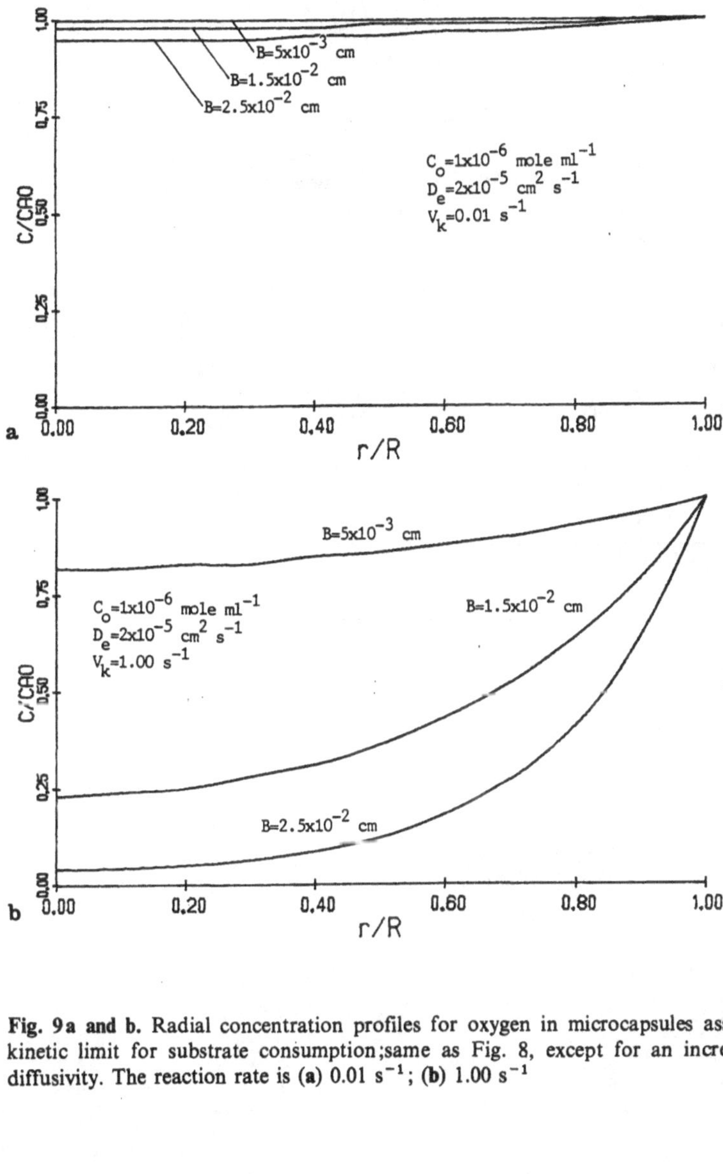

Fig. 9a and b. Radial concentration profiles for oxygen in microcapsules assuming a first order kinetic limit for substrate consumption;same as Fig. 8, except for an increase in the effective diffusivity. The reaction rate is **(a)** 0.01 s^{-1}; **(b)** 1.00 s^{-1}

Figure 11 demonstrates concentration profiles for glucose and oxygen assuming a first order limit to the rate equation. The axial concentration gradients shown in Fig. 11a are similar to those for the zero order kinetic limit. The radial concentration profiles, however, are drastically different. While the shape of the curves depends on the values of the kinetic constants used, the contrast between the radial concentration profiles for the two rate limits shows the importance of using the proper rate law and the applicable kinetic constants for a particular system. If the system is as depicted by Fig. 10, it should support cell growth, while if Fig. 11 is more appropriate, the system would most likely result in massive cell death.

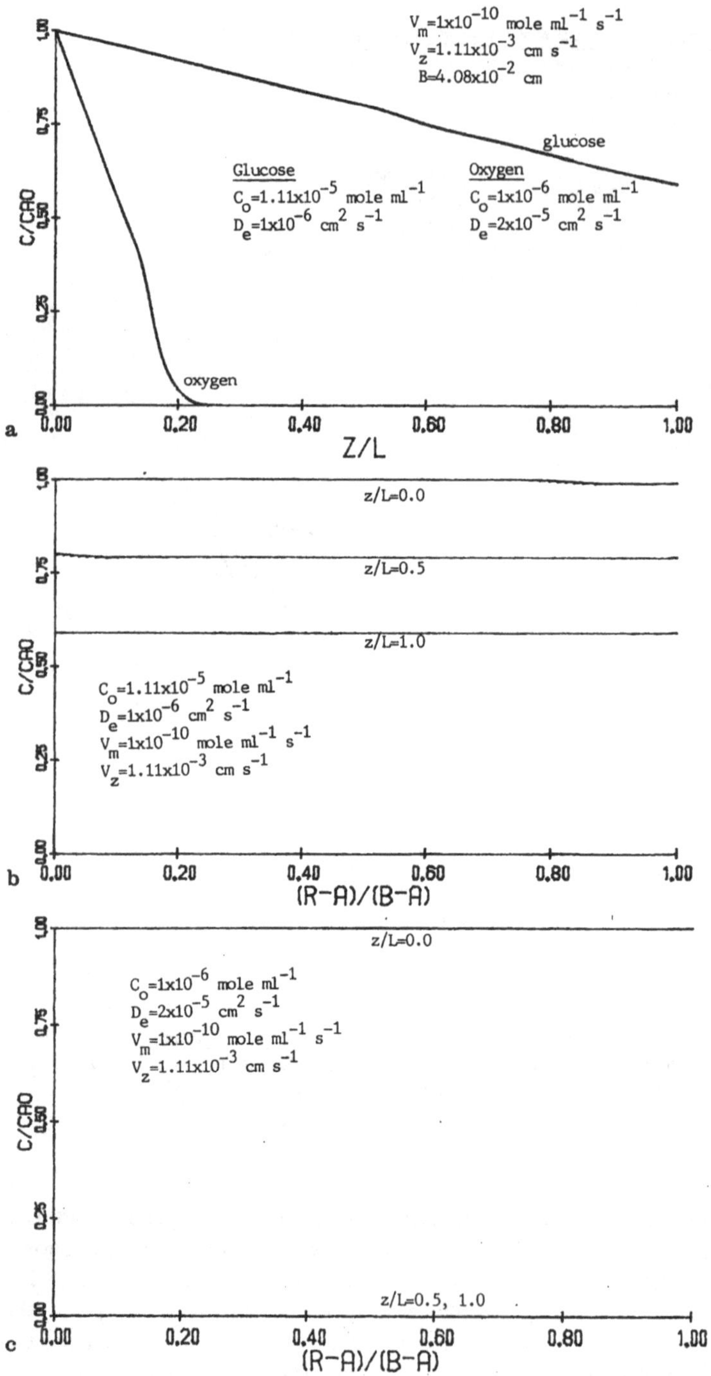

Fig. 10a–c. Concentration profiles for oxygen and glucose in hollow fiber reactors assuming a zero order kinetic limit for substrate consumption. **(a)** Axial concentration gradients of oxygen and glucose; **(b)** radial concentration gradients for glucose; **(c)** radial concentration gradients for oxygen

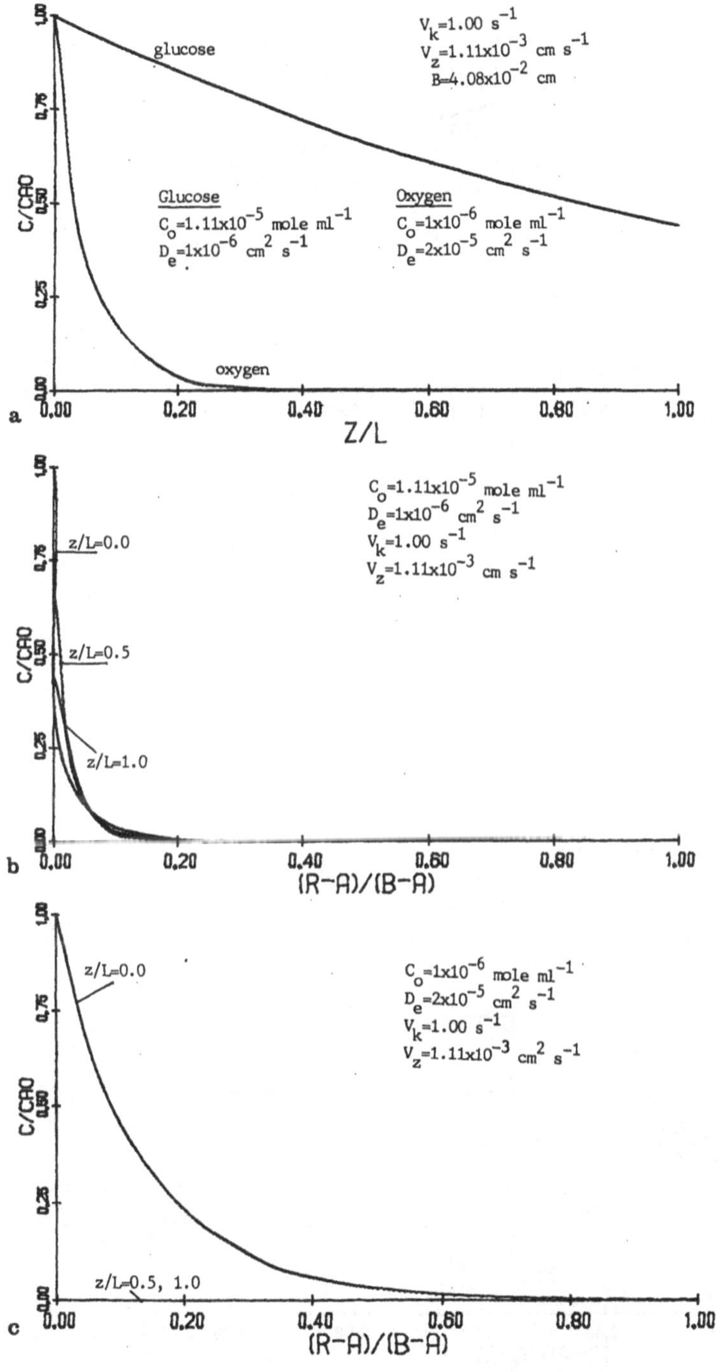

Fig. 11a–c. Concentration profiles for oxygen and glucose assuming a first order kinetic limit for substrate consumption. **(a)** Axial concentration gradients of oxygen and glucose; **(b)** radial concentration gradients for glucose; **(c)** radial concentration gradients for oxygen

a
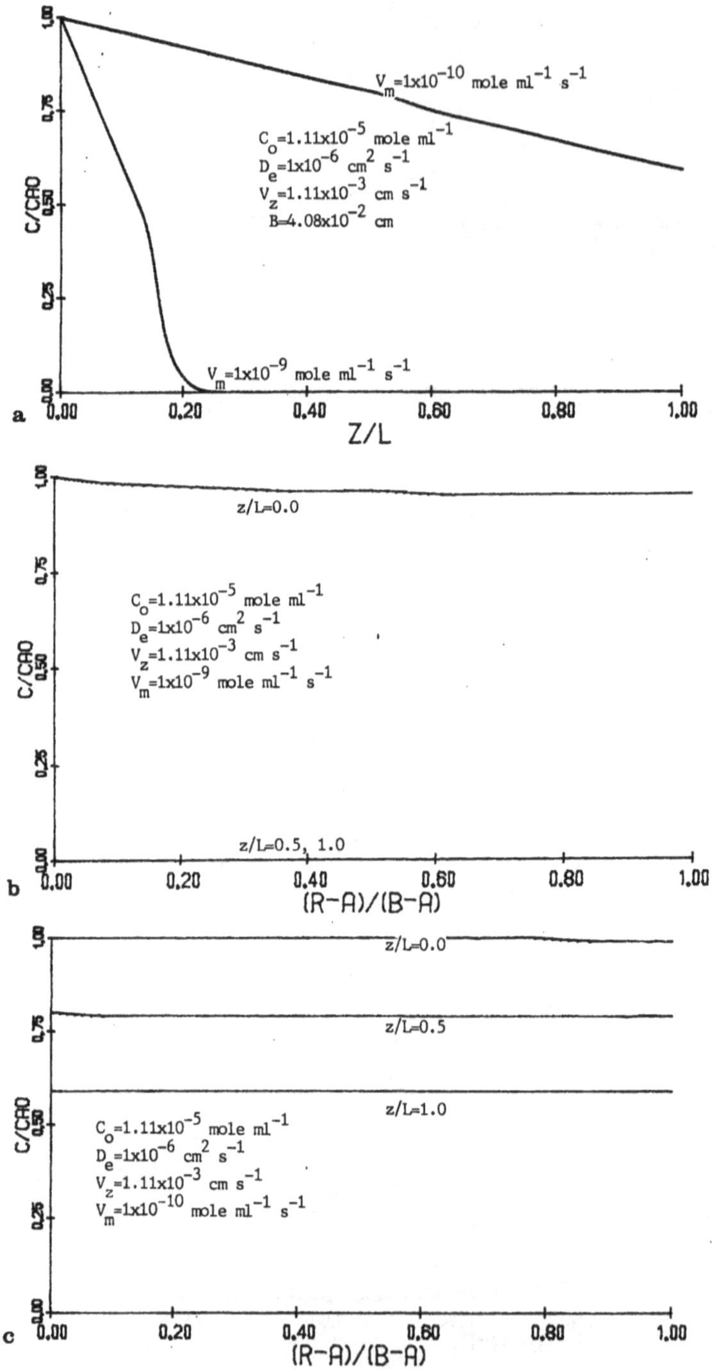

Fig. 12a–c. Concentration profiles for glucose in hollow fiber reactors assuming a zero order kinetic limit for substrate consumption. (**a**) Axial concentration gradients for two maximum reaction rates; (**b**) radial concentration gradients with $V_m = 1 \times 10^{-9}$ mole ml^{-1} s^{-1}; (**c**) radial concentration gradients with $V_m = 1 \times 10^{-10}$ mole ml^{-1} s^{-1}

Figure 12 portrays glucose concentration profiles for two maximum uptake rates, assuming a zero order limit to the rate equation. Simply decreasing the rate of substrate consumption by a factor of 10 converts an unfeasible system into a feasible one. The higher uptake rate effectively depletes the glucose from the fiber lumen. Again, as seen in Fig. 10, the potential for axial substrate diffusion is great, as seen by the radial concentration profiles at various distances along the reactor length. Despite the difference in axial concentration gradients, the radial concentration profiles all depict fairly flat curves. A slight change in slope is seen between Figs. 12b and 12c, but the difference isn't nearly as great as that seen in Fig. 12a. The reason for this is not known.

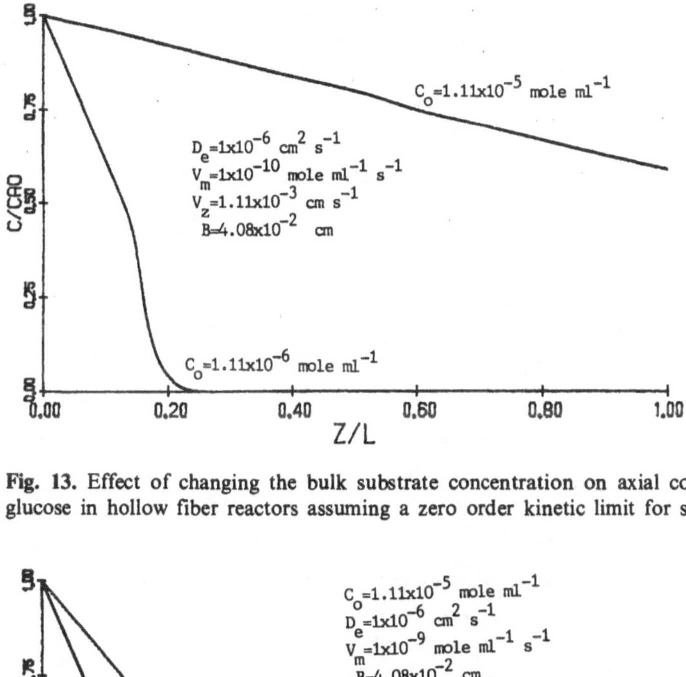

Fig. 13. Effect of changing the bulk substrate concentration on axial concentration profiles for glucose in hollow fiber reactors assuming a zero order kinetic limit for substrate consumption

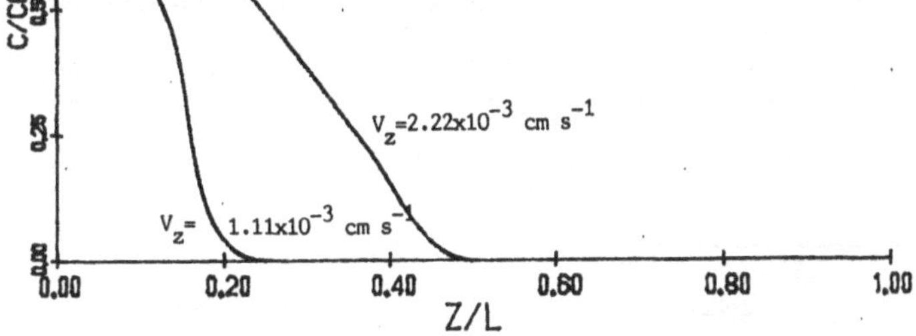

Fig. 14. Effect of varying axial velocity on concentration profiles of glucose in the hollow fiber lumen, assuming a zero order rate limit

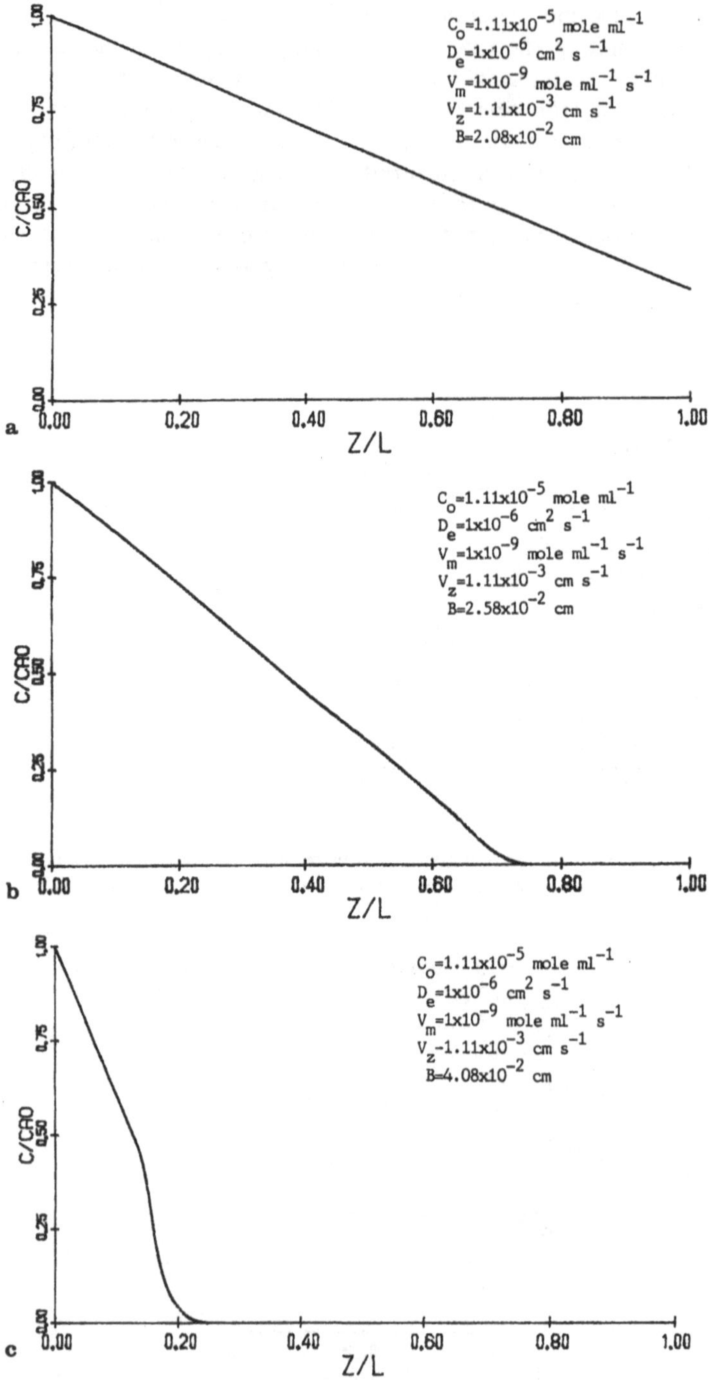

Fig. 15a–c. Glucose concentration profiles assuming a zero order kinetic limit for determining an effective radius of the annular region surrounding one hollow fiber membrane; **(a)** $B = 2.08 \times 10^{-2}$ cm; **(b)** $B = 2.58 \times 10^{-2}$ cm; **(c)** $B = 4.08 \times 10^{-2}$ cm (the actual reactor)

The effect of changing the bulk substrate concentration by a factor of 10 is shown in Fig. 13. The effect is the same as changing the maximum reaction rate by the same factor (see Fig. 12). The concentration profiles show the expected trend, that is, a higher bulk concentration results in higher concentration throughout the fiber lumen. It is important to note that the lower concentration exhibits a limiting behavior — it cannot supply sufficient glucose for the entire fiber length. Thus, a minimum acceptable substrate concentration, for these reactor conditions, exists between the two tested values.

When substrate concentration is limited, increasing the velocity of the media through the fiber lumens may improve the substrate availability in the reactor as shown in Fig. 14. Increasing the axial velocity by factor of 2 results in an approximate doubling of the reactor length that can be supplied with adequate nutrient. Thus, either substrate concentration or axial flow rate can be adjusted to accommodate the other when limiting conditions exist. For example, the concentration of oxygen in the media is limited by the solubility of oxygen in the media at a given temperature. To increase the availability of oxygen in the reactor, the media flow rate can be increased.

The model equations, besides yielding insight into resulting concentration profiles, can also be used to aid in hollow fiber reactor design. Figure 15 shows axial concentration profiles of glucose for the zero order kinetic limit using three different effective annulus radii. By finding the minimum radius of the region that can be supplied by an individual membrane, one can determine how many equally spaced fibers are required for a particular hollow fiber module. Since the radial concentration gradients are expected to be fairly flat for the zero order kinetic limit, only the axial concentration gradients are shown. As can be seen from the figure, only with an effective radius of approximately 2.08×10^{-2} cm (Fig. 15a) will there be enough glucose to supply cells along the entire length of the reactor. In both Figs. 15b and 15c, glucose is depleted prior to the reactor exit, leaving those cells at the further end without sufficient substrate. For this particular set of reactor conditions, the hollow fiber unit should contain four times as many fibers as it has in order to effectively supply the entire reactor volume. This procedure could also be used to determine an effective reactor length for a given number of membrane fibers.

The results for both the microcapsules and the hollow fiber reactors have demonstrated the variability of results obtained for different system parameters. This study has utilized several sources for the parameters used herein, some of which may not be appropriate for certain suspended mammalian cells. When working with a particular cell line, these parameters should be determined so that the reactor variables may then be manipulated to optimize the cellular environment. Simulations, such as the ones just presented, provide helpful insight toward the optimization process.

6 Discussion

The model equations presented in this paper cannot accurately represent the functioning bioreactor due to the assumptions made to simplify the governing equations. However, these equations, and others like them, can provide information leading to guidelines for reactor use and design. Coupled with experiments, model equations can yield some very useful insights.

The results have repeatedly shown the importance of obtaining accurate values of the substrate effective diffusivity and kinetic constants, and, to a lesser extent, the bulk concentration for the particular cell system being used. By changing the value of one of these parameters by an order of magnitude or less, the substrate concentration available to the cells may change from sufficient to vastly inadequate. Figure 2 demonstrates that an increase in the maximum reaction rate by a factor of 10 severely restricts the permissible microcapsule diameters. If microcapsule size is difficult to regulate, then this could pose a serious problem by decreasing productivity of cells in the large microcapsules. Another question is concerned with the minimum practical capsule size, i.e., how small can microcapsules be made, taking into account both the physical process and the economic considerations. Perhaps spheres with diameters of 100 microns ($B = 5 \times 10^{-3}$ cm), used in this study, are not feasible, even though they certainly will provide the most surface area for diffusion per unit volume. Glacken et al. [13] have calculated that, in order to support cell respiration, the capsule diameter would have to be 170 microns or less. This is a conservative estimate since the oxygen concentration that they used is very low. The point is that, for optimum production, the capsule size should be determined for every individual system.

Similar results were observed in the hollow fiber reactors following increases or decreases in parameter values, although the changes in gradient profiles were not as pronounced as for the microcapsules. It should be noted that the maximum rates of reaction used for the microcapsules were from 10 to 10^3 times as large as those used for the hollow fiber reactors. The latter could not support the higher uptake rates. While this difference may seem prohibitive, it does not preclude the use of hollow fiber reactors since many systems are regulated to minimize cell growth, following an initial growth period, and to maximize product generation. The resulting kinetic constants are certainly lower than those for maximum cell growth.

For either system, it is apparent that each of the required substrates must be investigated to determine the limiting factor. Concentration gradients for glucose may be adequate while those for oxygen may show depletion, limiting the rate of reaction and dictating the productive size of the reactor.

The above results are valid in general, yet they must be interpreted with consideration of the assumptions. For the microcapsules, it was assumed that there is no convective flux inside the spheres. With stirring, there will be some flow into and out of the capsule, the magnitude of which would depend on the rate of stirring, the capsule porosity, etc. With a flexible shell wall the contained cell suspension may not be stagnant as assumed and may exhibit some internal mixing, increasing the mass transfer rates. This should be investigated to see if it is indeed negligible. Another assumption was that the bulk concentration remained constant at its maximum value (media glucose concentration or oxygen saturation). While it is likely that a steady state is attained with a continuous flow system as long as there is a constant replenishment of old media, it is unlikely that this steady-state concentration is the maximum concentration, rather it will have a value lower than C_0. For a batch system or a recycle system without a continuous feed and bleed, the concentration of the bulk solution cannot be assumed to be a steady value. In this case, a transient term must be included in the model. This formulation would allow prediction of the optimum time intervals between media substitutions or additions.

The assumptions used in the hollow fiber system can be similarly explored. A prime target is the assumption that there is no diffusion in the axial direction. Investigation of the radial concentration gradients in Figs. 10, 11, and 12 indicate that axial diffusion is likely to occur. High substrate concentrations in the shell space at the entrance of the reactor would provide a driving force for axial diffusion which should result in radial concentration profiles at the reactor midpoint and beyond which are not quite as dismal as those indicated in the above-mentioned figures. Inclusion of an axial diffusion term into the continuity equation for extracapillary substrate concentrations does, however, create a more difficult problem which, most likely, can be solved by numerical techniques.

As with the microcapsules, it may be instructive to include a transient concentration term. The need for this depends on the hollow fiber system setup. If nutrients are supplied at a rate equal to their uptake, then no time dependence is required. However, if nutrients are added only periodically, then inclusion of a transient term may be necessary. As mentioned above, Webster and Shuler [40] have published a method for determining transient substrate concentration profiles in hollow fiber reactors. They determined that steady state profiles are established rapidly for systems with constant bulk substrate concentration, thus obviating the need for a transient term when single pass or continuous feed systems are employed.

Another possible source of inconsistency in the hollow fiber assumptions is that there is no convective flow in the annulus. The validity of this assumption depends on the pressure drop along the fibers: the fluid will travel along the path of least resistance. If a high pressure drop exists (most likely with larger reactors), fluid may enter the shell space through the fiber walls and flow amongst the cells, providing some mixing. The fluid will then reenter the fiber lumens near the end of the reactor, where the hydrostatic pressure gradient across the membrane is less. This has been called the Starling effect. A study by Wei and Russ [41] indicated that in hollow fiber reactors diffusion is the primary mode of transport for smaller molecules such as oxygen and glucose, especially when the cell mass is impermeable. As the cell mass becomes more permeable, convection plays an increasingly important role. With the low flow rates and short axial distance used in this paper, convective flow in the shell space is most likely negligible. However, it may be important to consider this form of transport in larger systems or in systems where cell redistribution and membrane fouling could be problems (causing large pressure drops from inlet to outlet). A mathematical analysis of convection superimposed onto diffusion has been developed for the first order rate limit by Schonberg and Belfort [32].

Membrane fouling itself may be a problem if high serum concentrations are used, if flow rates are too low, or if the membrane is too "tight". Generally, the problem of fouling can be minimized by periodically reversing or pulsing the flow. Choosing a membrane material with a low propensity to sorb proteins would help reduce the fouling problem. Other design considerations can also minimize fouling. If membrane fouling does occur to an appreciable extent then the membrane partition coefficient cannot be assumed to be one. Instead, there will be increased resistance to substrate diffusion especially beyond the entrance region. Considering the concentration profiles presented in the results section, this would cause further deterioration of the inadequate concentrations available to cells in the latter half of the reactor. Thus, membrane fouling should be avoided or kept to a minimum, if possible.

A topic which has not been considered is that of product removal. Certain cell byproducts, such as carbon dioxide, may cause feedback inhibition when their concentrations rise. In this case, the zero and first order limits of the Monod equation may not be acceptable. Accumulation of CO_2 gas within the sealed shell space could reduce or inhibit product formation and may even increase the total shell pressure creating defects in the hollow fiber and eventually rupturing the membranes [16]. Thus, byproduct removal is important for continuous systems.

The results obtained from the simulations lead to suggestions for minimizing and possibly overcoming some of the limitations of the system. Using microcapsules, it is important to use a sufficiently small capsule diameter, as determined by diffusional and kinetic parameters. It is also important to maintain a high bulk concentration of substrate to provide a sufficient driving force for diffusion.

With hollow fiber reactors, spatial arrangement of the fiber membranes is important. The simulation used in this report effectively assumed that the fibers are arranged symmetrically in the reactor so that the reactor volume can be divided into 150 small annular reactors, each with one membrane fiber in the center. As has been shown above, one can determine the volume of the annulus surrounding an individual membrane which can be supplied sufficiently with substrate. This type of analysis can suggest optimum fiber arrangements for a real system. It is important to effectively space the fibers, rather than clump them together in the core of the reactor. Since this may be difficult to regulate an efficiency factor could be incorporated into the design to account for areas within the reactor that may not be within the critical distance of a membrane. The critical distance is simply that distance beyond which no substrate exists. As was shown in several of the figures, reactor length may be a limiting factor. Based on the results given in this report, one could surmise that shorter modules could be a more effective means of supplying nutrients to cultured cells in hollow fiber reactors. No conclusion regarding this thought can be drawn until axial diffusion is considered.

If significant axial gradients truly exist in the hollow fiber modules, then cells near the reactor exit may not receive adequate substrate supply, while those near the reactor entrance will be surrounded by an overabundance of nutrients. This effect can be minimized by using higher flowrates or by periodically reversing the direction of axial flow. Both of these methods have been suggested by Hopkinson [14]. Another possibility to consider would be to use convective flow across the cell mass or to slowly move the cell mass itself, thereby reducing diffusional mass transfer gradients. With respect to oxygen, not only is the diffusion coefficient low in medium containing serum but the saturated partial pressure of oxygen at 37 °C in the medium is also very low. Besides convection, it may be possible to add "synthetic erythrocytes" or oxygen carrying capsules to the shell side, thereby increasing the available oxygen. The feasibility of this idea would have to be further investigated.

In conclusion, this analysis has provided a simplified simulation of substrate concentration profiles for microcapsules and a hollow fiber membrane reactor. From these profiles, generated by varying the parameters describing diffusion and kinetics, situations where diffusional limitations occur have been indentified. With this knowledge, steps can be taken to improve the substrate mass transfer and thus improve the efficiency of the reactor. Several suggestions were made toward this end for each reactor system.

7 Acknowledgment

The authors would like to thank Jeffrey Schonberg for his assistance.

8 Notation

A	radius of membrane fiber (cm)
B	microcapsule radius or hollow fiber annulus radius (cm)
C_0	bulk substrate concentration (moles ml^{-1})
C(r)	concentration as a function of radius (moles ml^{-1})
C(z)	concentration as a function of axial distance (moles ml^{-1})
D_e	effective diffusivity (cm^2 s^{-1})
$I_0(x)$	modified Bessel fuction of the first kind, order zero
$I_1(x)$	modified Bessel function of the first kind, order one
$K_0(x)$	modified Bessel function of the second kind, order zero
$K_1(x)$	modified Bessel function of the second kind, order one
K_m	Michaelis constant (moles ml^{-1})
L	hollow fiber length (cm)
n	number of hollow fibers
N	substrate flux (moles cm^2 s^{-1})
r	radius (cm)
R	outer radius of microcapsule or hollow fiber module (cm)
$\mathcal{R}$	uptake of substrate (moles ml^{-1} s^{-1})
V_k	V_m/K_m (s^{-1})
V_m	maximum reaction rate (moles ml^{-1} s^{-1})
V_z	axial fluid velocity (cm s^{-1})
z	axial distance (cm)

9 Appendix

Equation (14) is solved using modified Bessel functions [29].

$$\frac{d^2C}{dr^2} + \frac{1}{r}\frac{\partial C}{\partial r} \frac{V_m C}{K_m D_e} = 0 \qquad (14)$$

The solution is

$$C(r) = a\, I_0(xr) + b\, K_0(xr)$$

where

$$x = (V_m/D_e K_m)^{1/2} \qquad (24)$$

The constants, a and b, are determined from the boundary conditions

$$\text{at } r = A, C(A) = C(z) \quad (15a)$$

$$\text{at } r = B, \frac{dC}{dr} = 0 \quad (15b)$$

$$C(A) = C(z) = aI_0(xA) + bK_0(xA)$$

$$\frac{dC(B)}{dr} = aI_1(xB) - bK_1(xB)$$

These can be solved to yield

$$a = C(z)\, K_1(xB)/(K_0(xA)\, I_1(xB) + K_1(xB)\, I_0(xA))$$

$$b = C(z)\, I_1(xB)/(K_0(xA)\, I_1(xB) + K_1(xB)\, I_0(xA))$$

giving Eq. (23)

$$C(r) = C(z) \frac{K_1(xB)\, I_0(xr) + I_1(xB)\, K_0(xr)}{K_0(xA)\, I_1(xB) + K_1(xB)\, I_0(xA)} \quad (23)$$

10 References

1. Abramowitz, M., Stegun, I. A., (eds.): Handbook of Mathematical Functions. Dover, New York 1972
2. Amicon literature: 17 Cherry Hill Drive, Danvers, MA 01923 (1985)
3. Altshuler, G. L.: unpublished results (1985)
4. Altshuler, G. L., Dziewulski, D. M., Sowek, J. A., Belfort, G.: Biotech. Bioeng. *28*, 646 (1986)
5. Bailey, J. E., Ollis, D. F.: Biochemical Engineering Fundamentals. McGraw Hill, New York 1977
6. Bird, R. B., Stewart, W. E., Lightfoot, E. N.: Transport Phenomena, Wiley, New York 1960
7. Calabresi, P., McCarthy, K. L., Dexter, D. L., Cummings, F. J., Rotman, B.: Proc. Am. Assoc. Cancer Res. *22*, 302 (1981)
7a. David, G. S., Reisfeld, R. A., Chino, T. H.: J. Natl. Cancer Inst. *60*, 303 (1978)
8. Davis, M. E., Watson, L. T.: Biotech. Bioeng. *27*, 182 (1985)
9. Ehrlich, K. C., Stewart, E., Klein, E.: In Vitro *14*, 443 (1978)
10. Fleischaker, R. J., Giard, D. J., Weaver, J., Sinskey, A. J.: Adv. Biotechn. *1*, 425 (1980)
11. Freyer, J. P., Sutherland, R. M.: Adv. Exper. Med. Biol. *159*, 463 (1983)
12. Georgakis, C., Chan, P. C. H., Aris, R.: Biotech. Bioeng. *17*, 99 (1975)
13. Glacken, M. W., Fleischaker, R. J., Sinskey, A. J.: Ann. NY Acad. Sci. *413*, 355 (1983)
14. Hopkinson, J.: Bio/Technology *3*, 225 (1985)
15. Horvath, C., Shendalwan, L. H., Light, R. T.: Chem. Eng. Sci. *28*, 375 (1973)
16. Inloes, D. S., Taylor, D. P., Cohen, S. N., Michaels, A. S., Robertson, C. R.: Appl. Envir. Microb. *46*, 264 (1983)
17. Jarvis, A. P., Grdina, T. A.: Biotechniques *1*, 22 (1983)
18. Kan, J. K., Shuler, M. L.: Biotech. Bioeng. *20*, 217 (1978)
19. Karel, S. F., Libicki, S. B., Robertson, C. R.: Chem. Eng. Sci. *40*, 1321 (1985)
20. Kim S., Cooney, D. O.: Chem. Eng. Sci. *31*, 289 (1976)
21. Kleinstreuer, C., Poweigha, T.: Adv. Biochem. Eng./Biotech. *30*, 91 (1984)
22. Knazek, R. A., Gullino, P. M., Kohler, P. O., Dedrick, R. L.: Science *178*, 65 (1972)
23. Knazek, R. A., Kohler, P. O., Gullino, P. M.: Exp. Cell Res. *84*, 251 (1974)

23a. Knazek, R. A.: Fed. Proc. Am. Soc. Fed. Biol. *33*, 1978 (1979)
24. Ku, K., Kuo, M. J., Delente, J., Wildi, B. S., Feder, J.: Biotech. Bioeng. *23*, 79 (1981)
25. Lim, F., Moss, R. D.: J. Pharm. Sci. *70*, 351 (1981)
26. Lydersen, B. K., Pugh, G. G., Paris, M. S., Sharma, B. P., Noll, L. A.: Bio/Technology *3*, 63 (1985)
27. Margaritis, A., Wallace, J. B.: Bio/Technology *2*, 447 (1984)
28. Prenosil, J. E., Pedersen, H.: Enzyme Microb. Technol. *5*, 323 (1983)
28a. Ratner, P. L., Cleary, M. L., James E.: J. Virol. *26*, 536 (1978)
28b. Reisfeld, R. A., David, G. S., Ferrone, S., Pellegrino, M. A., Holmes, E. C.: Cancer Res. *37*, 2860 (1977)
29. Research and Education Assn.: Handbook of Mathematical Formulas, Tables, Functions, Graphs, Transforms. Research and Education Assn., New York 1980
30. Reuveny, S., Velez, D., Miller, L., MacMillan, D.: J. Immunol. Meth., in press (1986)
31. Rony, P. R.: Biotech. Bioeng. *13*, 431 (1971)
31a. Rutzky, L. P., Tomita, J. T., Calenoff, M. A., Kahan, B. D.: J. Natl. Cancer Inst. *63*, 893 (1979)
32. Schonberg, J. A., Belfort, G.: Enhanced nutrient transport in hollow fiber bioreactors: A theoretical analysis. Submitted (1986)
33. Sowek, J.: private communication (1985)
34. Tharakan, J. P., Chau, P. C.: Biotech. Bioeng. *28*, 329 (1986)
35. van Heuven, J. W., van Maanen, H. C. H. J., Ligtermoet, R.: Characteristics of immobilized enzyme systems. In: Innovations in Biotechnology, (Houwink, E. H. and van der Meer, R. R., eds.) p. 53. Elsevier Science, Amsterdam 1984
36. Waterland, L. R., Michaels, A. S., Robertson, C. R.: AIChE J. *20*, 50 (1974)
37. Waterland, L. R., Robertson, C. R., Michaels, A. S.: Chem. Eng. Commun. *2*, 37 (1975)
38. Webster, I. A., Shuler, M. L.: Biotech. Bioeng. *20*, 1541 (1978)
39. Webster, I. A., Shuler, M. L., Rony, P. R.: Biotech. Bioeng. *21*, 1725 (1979)
40. Webster, I. A., Shuler, M. L.: Biotech. Bioeng. *23*, 447 (1981)
41. Wei, J., Russ, M. B.: J. Theor. Biol. *66*, 775 (1977)
42. Wiemann, M. C., Ball, E. D., Fanger, M. W., Dexter, D. L., McIntyre, O. R., Bernier, G., Calabresi, P.: Clin. Res. *31*, 511A (1983)
43. Wolf, C. F. W., Munkelt, B. E.: Trans. Am. Soc. Artif. Int. Organs *21*, 16 (1975)

Nerve and Muscle Cells on Microcarriers in Culture

A. Shahar and S. Reuveny

Section of Electron Microscopy, Department of Biotechnology, Israel Institute for Biological Research, P.O. Box 19, Ness-Ziona 70450, Israel

Commercially available microcarriers (MCs), which are briefly described here, are usually used as a support for culturing a variety of cells both on laboratory and on an industrial scale. Selected MCs are used as a culture system for embryonic dissociated CNS cells and myoblasts. The tridimensional support provided by the MCs, enables neuronal and muscle cells to grow and differentiate into functional cell-MC units, which remain floating in the culture medium. The neuronal entities are characterized by intensive fiber growth followed by synaptogenesis and myelination. The muscular units develop striated myotubes, having the same orientation, which contract spontaneously. The MC technique is advantageous over the conventional monolayer procedure, allowing cells to grow to higher amounts at a better efficiency for longer periods. Functional units can be sampled without interfering with the ongoing culture. We consider the use of these cultures, as well as nerve muscle MC co-cultures as a tool for the study of neurotoxicology. The possibility of implantation of such functional culture units into injured adult nerve and muscle tissues is being considered.

Abbreviations

AChE Acetylcholinesterase
CNS Central nervous system
DIV Days in vitro
MC Microcarrier
MCs Microcarriers
SEM Scanning electron microscope or micrographs
TEM Transmission electron microscope or micrographs

1 Introduction

The development of microcarrier (MC) technology in tissue culture originated from the need for mass cultivation of mammalian cells in order to produce commercially biological cell products like viral vaccine, beta interferon etc. The concept of the microcarrier culturing system was first introduced by van Wezel in 1967. In this method cells were propagated on the surface of small solid particles suspended in the growth medium by slow agitation. The cells attached to the MC and grew gradually up to confluency on their surface.

Several cultivation systems were used for propagating cells on MCs. On a laboratory scale, cells were propagated on MCs in stationary Petri dishes, shake flasks, roller bottles and spinner flasks. On an industrial scale, cells were cultivated in fermentors up to 1000 liters volume. A wide range of cells from invertebrate, fish, bird and mammalian origin, transformed and normal cells, cell lines and lately genetically modified cells, have been cultivated on a variety of MCs [1].

The main advantages of this technology over monolayer cultures are: Availability of high surface to volume ratio which can be varied by changing the MC concentration; cells grow in a homogeneous culture in which environmental conditions are monitored and controlled; representative sampling is easily performed during cultivation; cell propagation can be carried out in a single high productivity vessel, rather than using many small low volume units; and the system can be relatively easily scaled up. A detailed description of the microcarrier technology and its advantages has been presented by Reuveny [1]. As mentioned previously, MCs are used mainly for mass cultivation of cells for production of viral vaccines and biologicals. However, recently, MCs are found to be an important tool in cell culture research, with various kinds of applications. Several laboratories [2,3,4] have shown that cells can be sub-cultivated by transfer from MCs to the dish without application of proteolytic enzymes. Duda [5] has shown that cells can be cryopreserved, while attached to MCs. Microcarrier cultures are processed for transmission (TEM) and scanning (SEM) electron microscopy [6,7]. Crespi and Thilly [8] have used synchronized mitotic cell population to study cell cycle dependence of certain mutagenic chemicals. Cells on MC are used in membrane isolation and membrane research [9,10,11], and for studies of cell-cell and cell-substratum interactions [12,13]. Kotler et al. [14] have shown that non-transformed cells which grow in dishes as monolayers can grow on MCs in multilayers. They suggested that the ability of cells to grow in multilayers is determined not only by their state of transformation but also by the type of support they are growing on. Davis and Kerr [15] and Davis et al. [16] have co-cultivated vascular endothelial cells and smooth muscle cells, using MCs, in order to study interaction between these two types of cells. Ren [17] used MCs for depleting macrophages from mouse spleen cell population.

The ability to propagate differentiating cells at high densities in a three dimensional mode on MCs rather than as bidimensional monolayers, provides a unique opportunity for studying cell differentiation, interaction and function. Differentiating bone cells [18], thyroid cells [19], pituitary cells [20,21], pancreatic beta-cells [22], endothelial cells [23], chick embryo muscle cells [24,25] and mouse peritoneal macrophages [26], were propagated on cylindrical and beaded MCs. These cells were propagated in batch cultures mainly for morphological studies, or in constant environment perfusion cultures [22, 23, 27], mainly for pharmacological studies.

Primary cultures of dissociated central nervous system (CNS), striated muscle and nerve-muscle co-cultures were sucessfully grown on MCs [6, 24, 28–31]. The rationale behind introducing MCs to primary cultures of neurons and muscle cells, was to provide these cells with a tridimensional support, enabling them to exhibit a growth pattern, close to the in vivo situation.

We review here neuronal and muscular differentiation achieved with conventional methods in comparison with the MCs technique. Further prospectives and possible applications of CNS and striated muscle MC cultures are discussed.

2 Commercial Microcarriers

Several kinds of MCs are now commercially available from a variety of sources. These microcarriers are made of hydrophilic polymers; dextran and cellulose, which are porous in their nature or polystyrene and glass which are non porous. Most of the MCs have a beaded form (only the cellulose microcarriers are cylindrical). They are 100–200 μm in size, have a smooth surface and a specific density of 1.03–1.1. The majority of the MCs are transparent to allow microscopic observation of the cells on them. A list of the commercially available MCs and their properties is given in Table 1. These microcarriers can be divided into 6 main groups.

1) Tertiary amino derivatized microcarriers (Cytodex 1, Super-beads). These are beaded MCs having a porous, non-rigid hydrophilic matrix (dextran) with positively charged tertiary amino groups, distributed throughout the whole matrix of the MCs. Since cells from animal and human source possess a net negative charge on their surface at a physiological pH, there is electrostatical attraction of the cells to the positively charged MCs. Various components of the growth media can penetrate into the porous microcarriers and are also attracted to the positively charged groups.

2) Surface charged microcarriers (Cytodex 2). These are quartenary amine-derivatized beaded MCs in which the charged groups are distributed in a layer on the outer surface of the microcarrier in order to reduce adsorption of growth substance into the MCs.

3) Collagen coated microcarriers or gelatin microcarriers (Cytodex 3, Ventragel, Gelibeads). These are beaded MCs which are either dextran beads covered with a layer of covalently bound denaturated collagen (Cytodex 3), or composed entirely of cross linked gelatin (Ventragel, Gelibeads). Cells attach onto the denaturated collagen. Since these MCs are non-charged, they absorb less serum protein from the culture medium as compared with positively charged microcarriers.

4) Polystyrene microcarriers (Biosilon, Cytospheres). These microcarriers have a hydrophobic non porous matrix (tissue culture treated polystyrene) with a low negative charge on their surface. Since these MCs are non porous there is no adsorption of growth medium into the microcarrier matrix and the microcarrier does not shrink during cell culture preparation for microscopy.

5) Glass microcarriers (Bioglass). These MCs are glass coated beads. Similar to the polystyrene microcarriers, they are non porous in nature. The exact nature of the coating is not specified.

Table 1. Commercial microcarrier

Reg. trade name and manufr.	Main property	Charged groups and exchange cap. or equivalent	Matrix composition
Cytodex 1 Pharmacia, Sweden	Positively charged groups distributed throughout the MC matrix	Diethylaminoethyl (DEAE) 1.5 meq g^{-1} dry materials	Dextran
Superbeads Flow Labs. U.S.A.	Positively charged groups distributed throughout the matrix	Diethylaminoethyl (DEAE) 2.0 meq g^{-1} dry dextran	Dextran
Cytodex 2 Pharmacia, Sweden	Positively charged groups distributed in a layer on the outer surface of the MC	Trimethyl-2-hydroxy-aminopropyl 0.6 meq g^{-1} dry materials	
Cytodex 3 Pharmacia, Sweden	Type 1 denaturated collagen covalently linked to the outer surface of the MC	60 µg collagen per cm^2 MC surface	Dextran
Ventragel Ventrex, U.S.A.	Cross-linked gelatin (denaturated collagen) MC	Gelatin	Cross-linked gelatin
Gelibeads KC Biological, U.S.A.	Cross-linked gelatin (denaturated collagen) MC	Gelatin	Cross-linked gelatin
Biosilon Nunc. Denmark	Tissue culture treated Polystyrene MC	Negative charges (tissue culture treatment)	Polystyrene
Cytospheres Lux, U.S.A.	Tissue culture treated polystyrene MC	Negative charges (tissue culture treatment)	Polystyrene
Bioglas Solohill Engineering Inc. U.S.A.	Glass coated beads	Not specified	Glass
DE-53 Whatman, U.K.	Positively charged groups distributed throughout the cylindrical cellulose matrix	DEAE 2.0 meq g^{-1} dry materials	Microgranular cellulose

N.D. — No data available

6) DEAE-cellulose microcarriers (DE-53). These are cylindrical shaped MCs having a microcrystalline cellulose matrix and charged with tertiary amine groups throughout the matrix. DE-53 is produced by Whatman (England) as an anion exchange resin for chromatography. However, Reuveny et al.[32,33], have found that this product can be used as MCs for cell cultivation. These MCs are unique in their cylindrical shape which allow higher surface/volume ratio as compared with the beaded microcarriers. Moreover, cells, especially primary cells or cells from diploid cell strains, tend to grow on these MCs to form multilayered aggregates composed of cells and MCs.

Shape and dimensions (diam. μm)	Surface area ($cm^2 g^{-1}$ dry weight MC)	Specific gravity	Porosity	Transparency	Ref.
Beads 131–210 mm	6000	1.03	+	+	[34]
Beads 135–205 mm	5000–6000	N.D.	+	+	[35]
Beads 114–198 mm	5500	1.04	+	+	[36]
Beads 133–215 mm	4600	1.04	+	+	[36]
Beads 150–250 mm	N.D.	N.D.	+	+	
Beads 115–235 mm	3300–4300	1.03–1.04	+	+	[37]
Beads 160–300 mm	225	1.05	−	+	[38, 39]
Beads 160–300 mm	250	1.04	−	±	
Beads 90–150 mm or 150–210 mm	350	1.03 or 1.04	−	±	[40, 41]
Cylinders 40–50 mm 80–400 mm length	N.D.	N.D.	+	−	[32, 33]

3 Primary Cultures of the Central Nervous System

3.1 Explants and Slices

Neurons of the central nervous system (CNS) have an organotypic organization in the brain and in the spinal cord forming structures of overlapping layers or grouped nuclei. Such a degree of tissue organization is simulated in vitro by organotypic cultures composed of explants (1 mm^3) or slices (400 μ) which grow in culture retaining most of their original structural relationship [42,43]. Because of their thickness, organotypic cultures usually do not allow a clear visualization of single living neurons. However, some cells can be identified on the basis of their size, location and arborization [44]. The organotypic cultures maintain their functional aspects for long periods

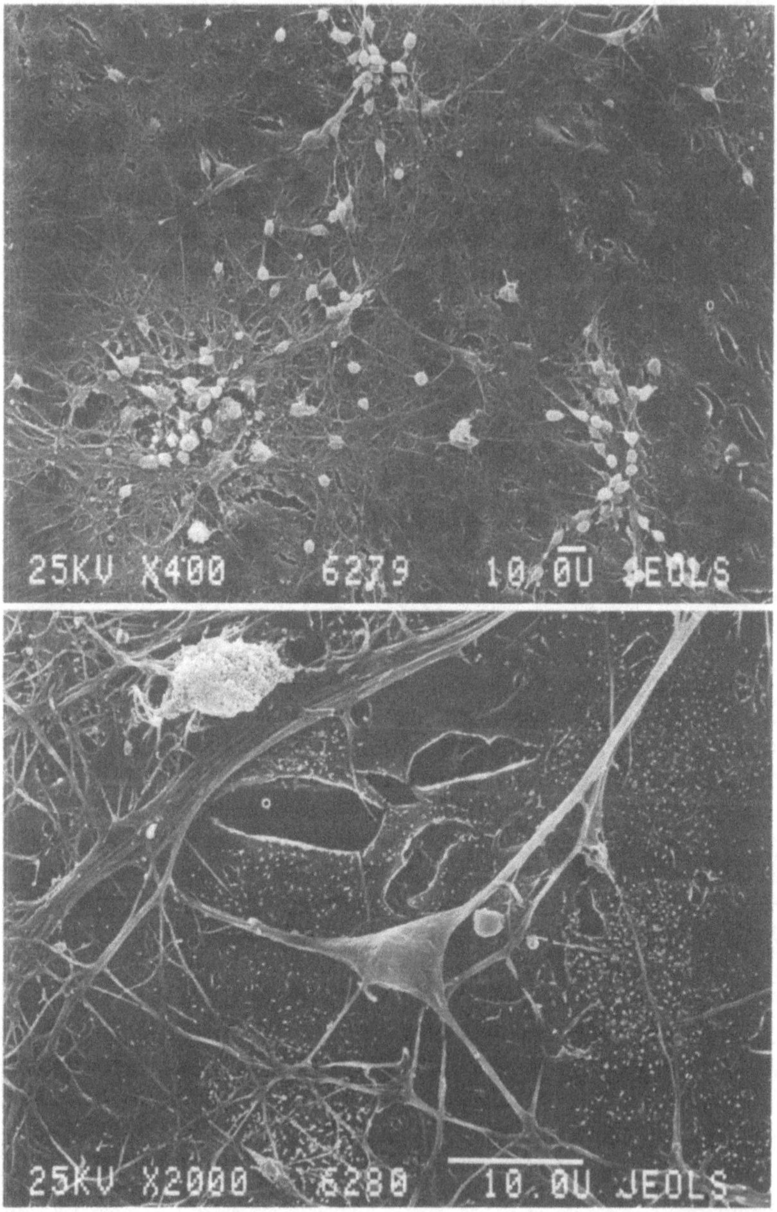

a

b

Fig. 1a and b. Scanning electron micrographs (SEM) of dissociated cerebral neurons. 10 d in vitro (DIV). Note: Perikarya with sprouting of fibers above a layer of non-neuronal elements exhibiting many microvilli. Rat fetuses dissociated brain 7 DIV

(2–4 months). They reach an advanced stage of maturation and myelinate extensively. The introduction of standardized slicing technique has significantly reduced variations between cultures of the same anatomical origin and has improved reproducibility [44].

3.2 Dissociated Cells

In cultivating dissociated CNS cells, the in vivo organization is disrupted during the dissociation procedure and is re-established only to a certain degree during cultivation. Dissociated cells are cultured either on a coated surface as monolayers [45, 46, 47, 48], or in rotating systems which allow the cells to form floating aggregates [49, 50].

In a monolayer culture system, single or small groups of neurons are randomly attached and sprout above a layer of dividing non-neuronal elements (Fig. 1a, b). These conditions do not allow nerve cells to achieve sufficient cell differentiation and organotypic tissue organization to simulate the in vivo neuronal structure. On the other hand, neurons can be continuously observed and are highly accessible for morphological immunocytochemical and electrophysiological studies. In aggregating CNS cultures neuronal elements rearrange themselves in a tridimensional structure, which facilitates cell contact, cell recognition and sorting out of cells. Thus, in a way, these cultures resemble organotypic cultures in their reorganization pattern and their ability to reach an advanced stage of maturation and myelination [51]. However, when the aggregate has reached a certain size, the compact growth of its cells at the periphery of the aggregate prevents optimal nutrition from the cells in the center leading to degenerative phenomena [52].

4 Cultures of Dissociated Central Nervous System on Microcarriers

The introduction of MCs to neuronal cultures was mainly for the purpose of providing the cells with a tridimensional anchorage. Growth and development of cells on this support took place on one or more layers, creating an environment allowing penetration of nutrient media favoring differentiation and myelination.

4.1 Cultures on Spherical Microcarriers

The first attempts to grow dissociated CNS cells on MCs were made with spheric beads [6]. Due to the transparency of these beads, the attached cells could be identified on the MCs by phase-contrast microscopy (Fig. 2a). Almost the same fraction of cells (about 90%) attached to the tertiary amino derivatized MCs (Cytodex 1), surface charged MCs (Cytodex 2) or Collagen coated MCs (Cytodex 3), see Sect. 2. Already during the first hour after seeding, there was a tendency for the beads, bearing the cells, to aggregate into clusters, each composed of several MCs. Such cells — MCs conglomerates, became more tightly interconnected by a ramified network of neuronal fibers sprouting during the first week of cultivation. SEM showed that in many beads, the fiber network had retracted, leaving a large part of the bead's surface uncovered

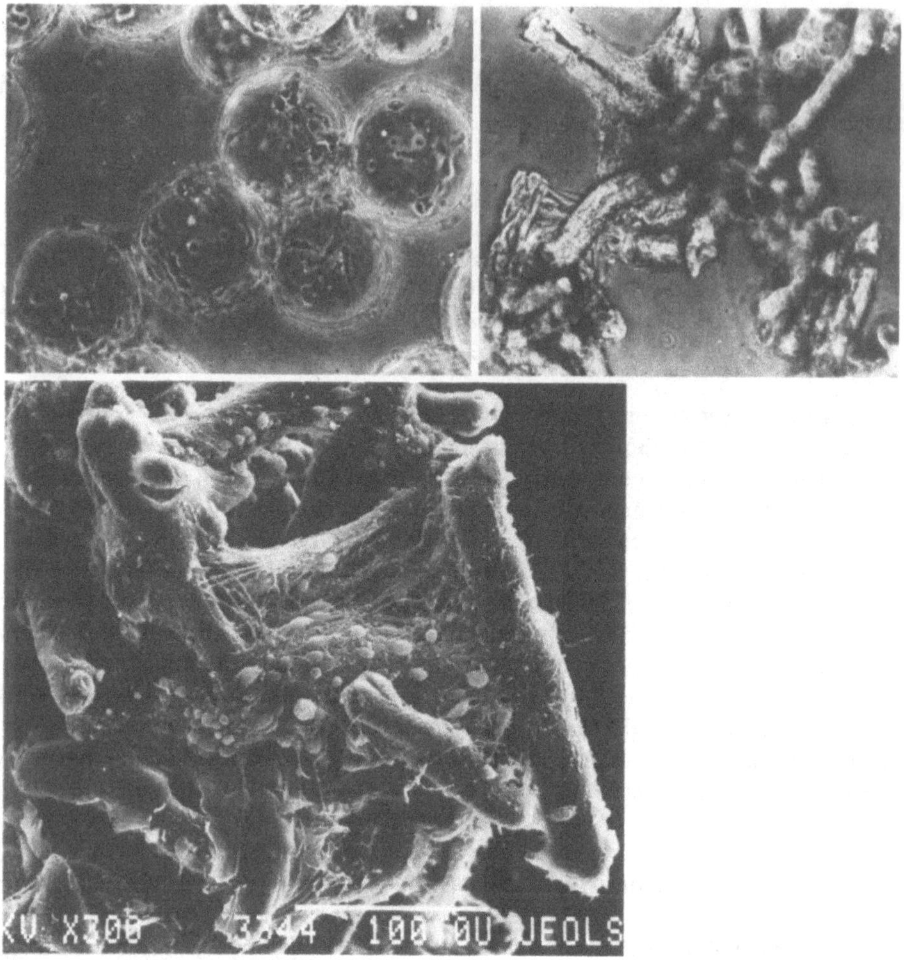

Fig. 2a–c. Phase contrast micrographs of single neurons on transparent Cytodex-1 beads in **a** and cell aggregation on DE-53 cylindrical MC after 4 d in culture in **b** SEM of a cell-MC conglomerate is shown in **c**

(Fig. 3a). In a few beads however, the surface was covered only by single coiled interconnected fibers (Fig. 3b). The unusual growth of thin coiled fibers, as well as the retraction of the developed fiber network, indicated that there was no firm attachment of the growing mass to the surface of the beads. This assumption is supported by the fact that most of the cells detach from the beads during washes in their processing for biochemical or ultra-structural analysis.

4.2 Cultures on Cylindrical Microcarriers

Cylindrical MCs (DE-53 Pre-swollen microgranular DEAE cellulose union exchange, Whatman) were found to be more suitable for neuronal growth than the spherical (Cytodex-1, Pharmacia) beads [6,29]. Although they are not transparent, masses of

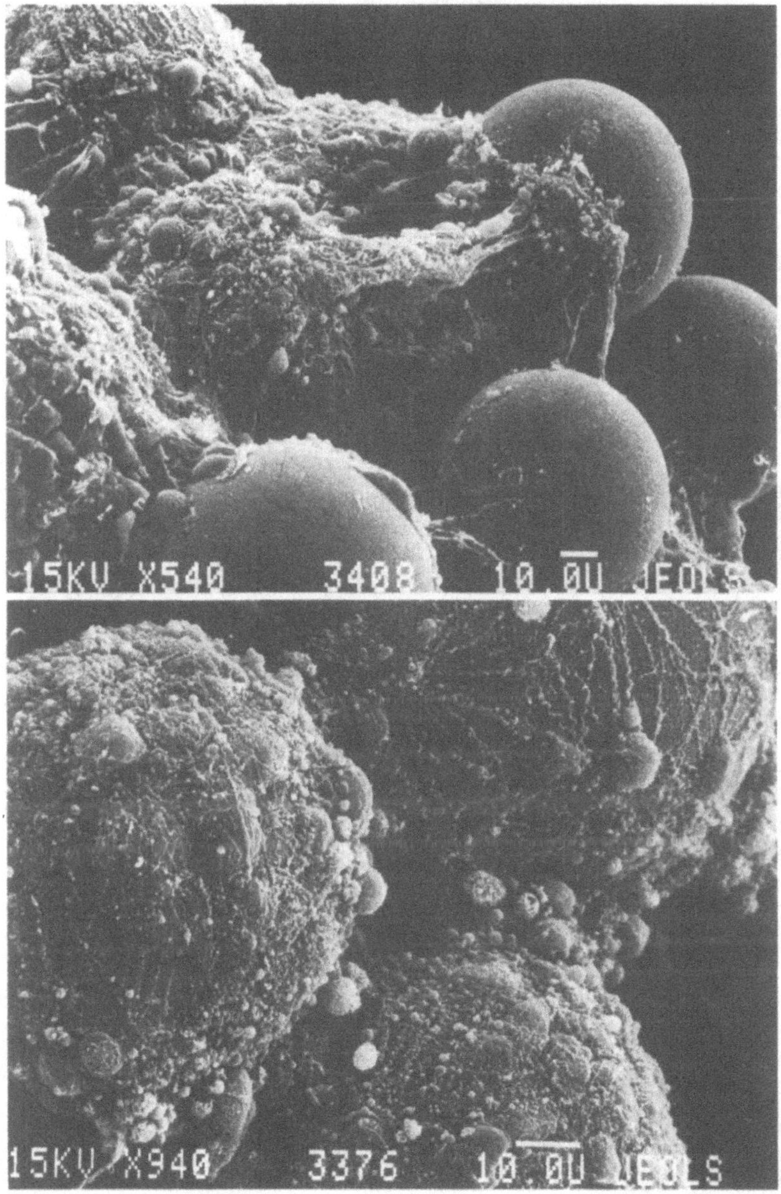

a

b

Fig. 3a and b. Stationary cultures of dissociated brain cells from 16 d old rat fetus 10 DIV. SEM showing in **a** retracted cell outgrowth which covers the beads only partially; **b** growth of single coiled fibers

cells and bundles of interconnecting fibers could be recognized in phase-contrast microscopy (Fig. 2b). In contrast to the fiber growth pattern described for the spherical beads, the cells on DE-53 conglomerates showed an intensive fiber growth which developed into a well-organized and established net-work. In SEM, it appeared like

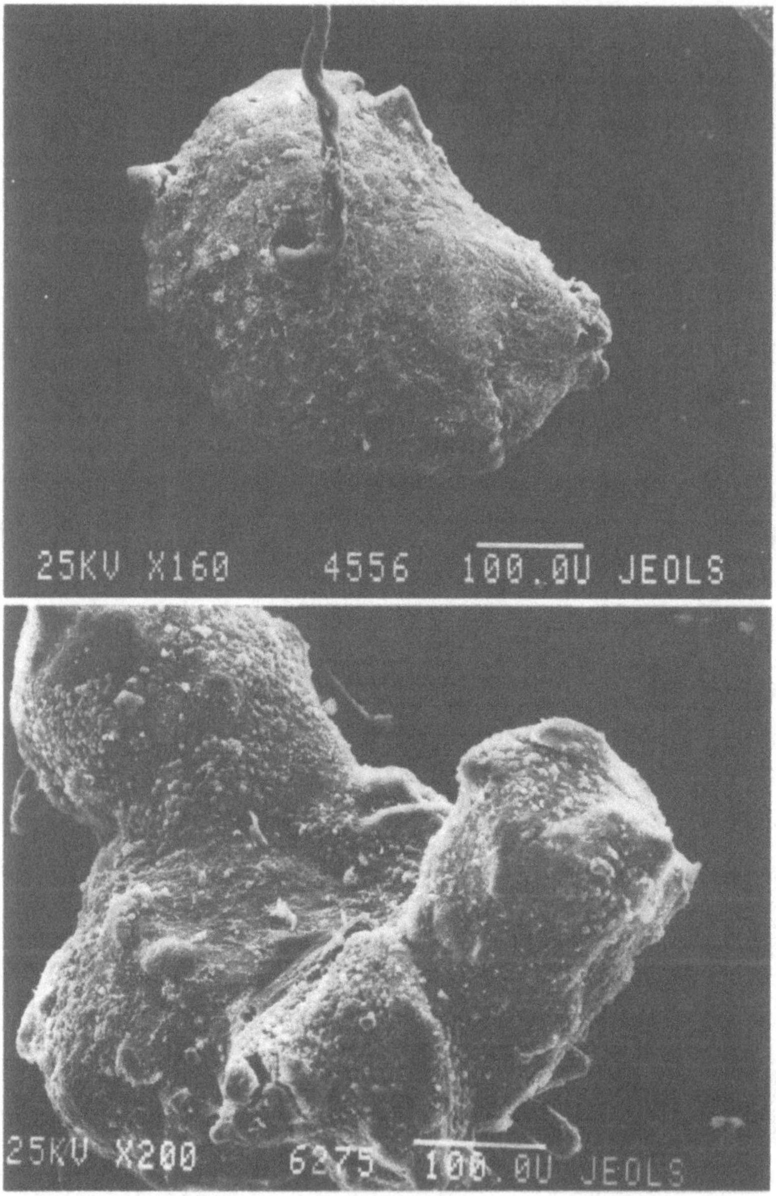

a

b

Fig. 4a and b. Brain cells dissociated from rat fetuses and cultured for 10 d on DE-53 cylindrical MCs. SEM of two conglomerates. **a** in the form of a wool yarn; **b** showing areas with numerous perikarya

a ball of wool yarn, firmly attached to the MCs and covering almost their entire surface (Figs. 2c, 4a). In several areas among the fibers, numerous oval smooth perikarya were observed (Fig. 4b).

An interesting observation is the appearance in culture of conglomerates covered by flat cells instead of the classical form of a wool yarn ball. The appearance of this

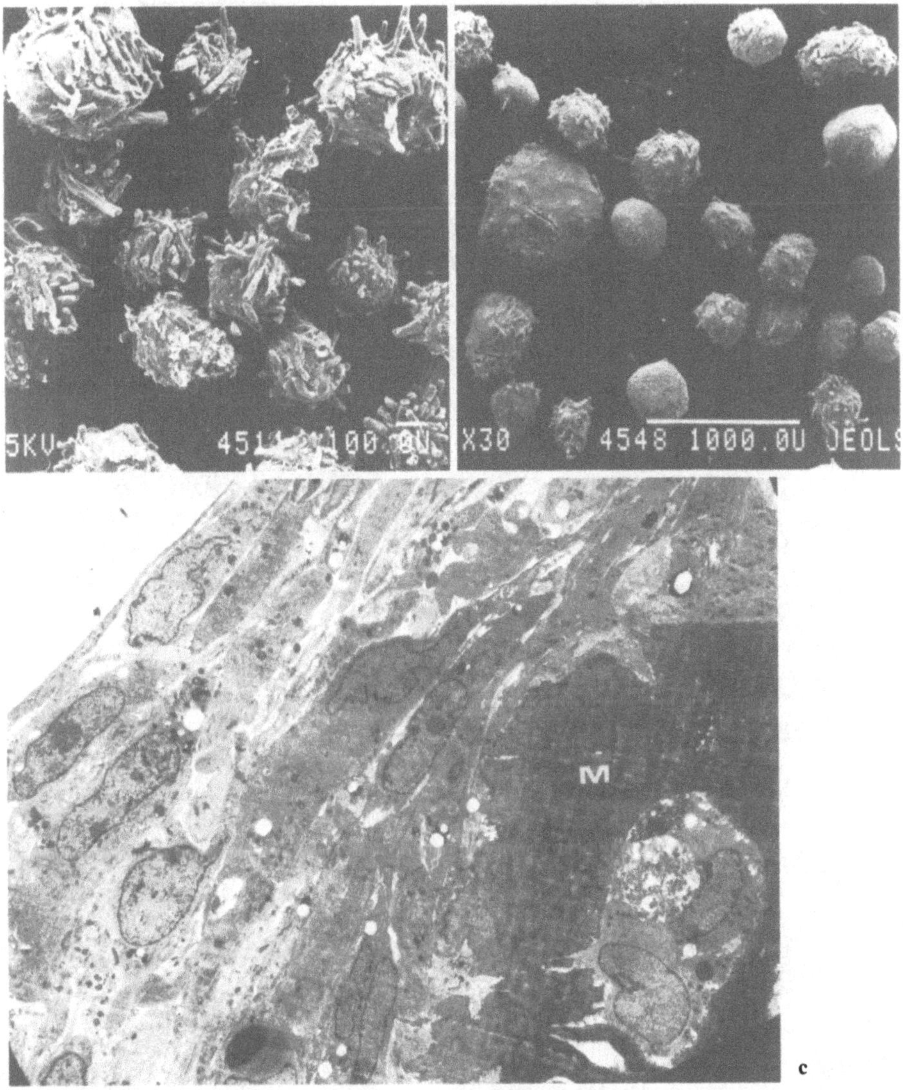

a b c

Fig. 5a–c. Dissociated spinal cord from 16 d rat fetus **a**, **b**, SEM of conglomerates entirely composed of flat cells. **c**, Overlapping layers of flat cells. Note: cellular growth within a hole in a MC × 2500

type of conglomerate was noticed more often in spinal cord, than in cerebral MC cultures. This is probably becausc in these cultures, part of the meninges remained attached to the dorsal root ganglia which are to be used in the cultures. In TEM the flat cells were arranged in several overlapping layers (Fig. 5c). In SEM, the adjacent cells of the external layer exhibited numerous microvilli (Fig. 5a, b). The fact that neuronal growth was not observed in these conglomerates might be due to malnutrition of cells in the center of the conglomerate, caused by the sealing coat of the flat cells. In addition to the regenerated fibers net work, the spaces of the conglomerates

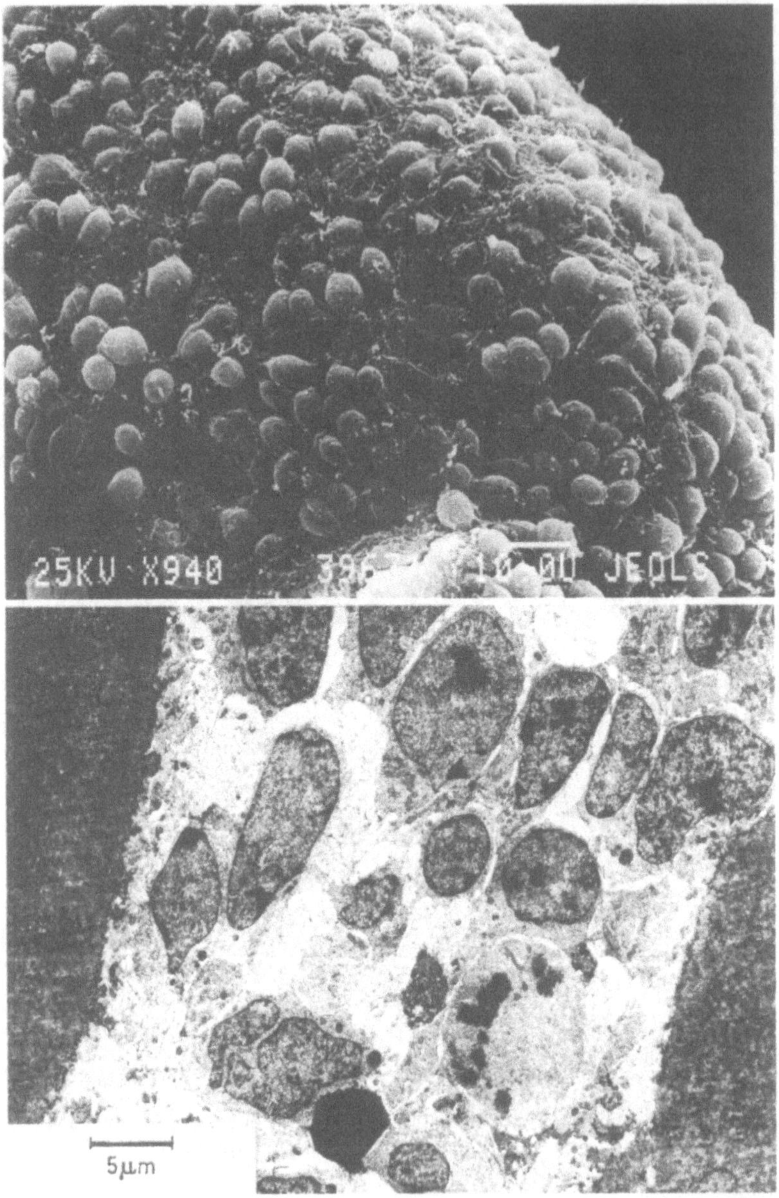

a

b

Fig. 6a and b. a, SEM of an area of aggregated perikarya from cerebral rat brain, 10 DIV; **b,** TEM (transmission electron micrograph) of rat fetus cerebral cells, 4 DIV, arranged between two MCs. Note: heterochromatic nuclei and a dividing cell

were filled up by many cells, usually gathered up in clumps. These cells, which at this time, could not be identified as neurons or glia elements, arranged themselves either tightly along the MCs or within small holes in their surface (Figs. 5c, 6a, b). The fact that neuronal tissue could grow and differentiate in such a close vicinity to the MC surface, ruled out the possibility of toxicity of the MC to these cells.

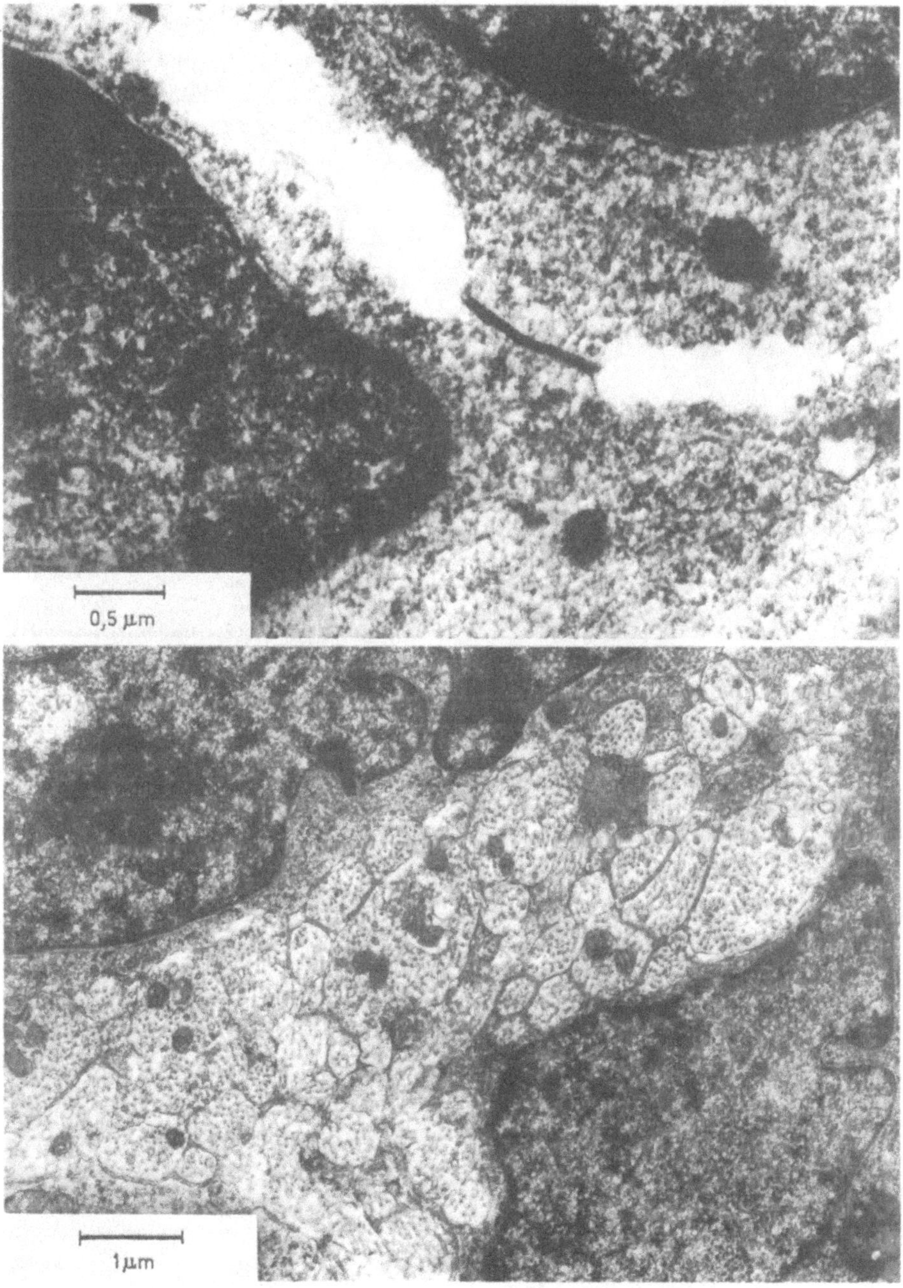

a

b

Fig. 7 a and b. TEM of a 4 d culture of dissociated brain from rat fetus. **a** Junction; **b** bundle of naked fibers

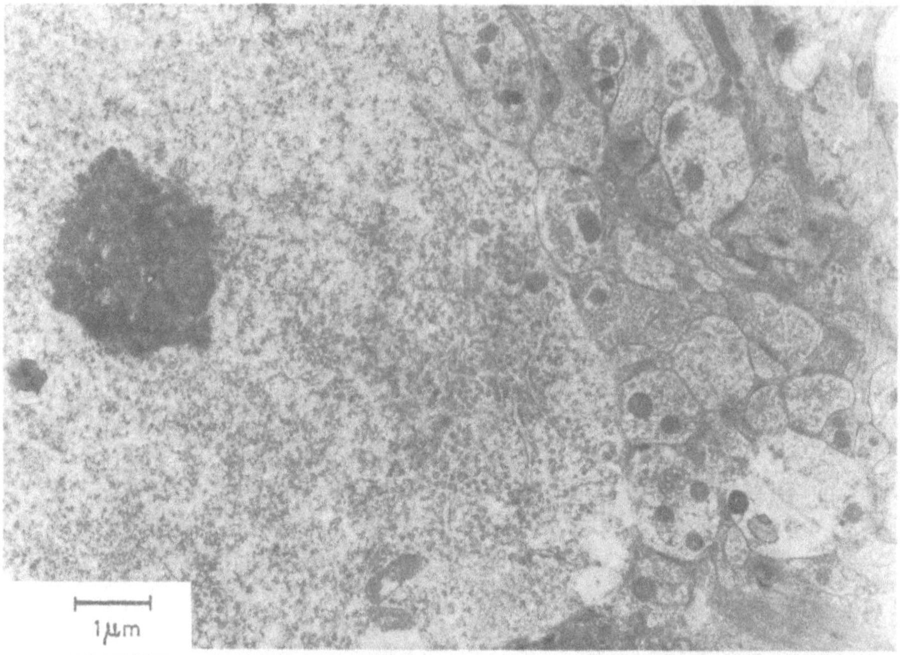

a

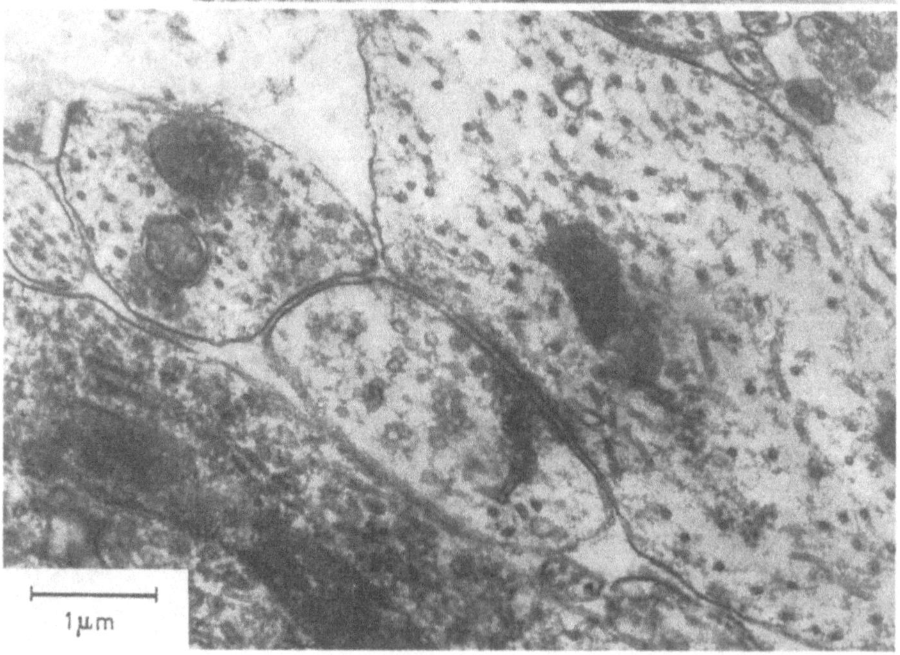

b

Fig. 8a and b. TEM of spinal cord MC cultures from 15 d rat fetus. **a** Part of a perikaryon contornated by many synaptic buttons; **b** a mature synapse

During the first week in culture in TEM scattered mitotic figures were evident (Fig. 6b). In cerebral cultures, most of the cells contained large heterochromatic nuclei with one or more nucleoli. The majority of spinal cord cells exhibited prominent euchromatic nuclei which resembled those of mature neurons.

Cells appeared to adhere to each other at different sites of their membrane by tight junction-like formations (Fig. 7a). Similar "Punta adherence" junctions were described at the same age, in aggregating cultures [49]. These points of contact did not exhibit a typical desmosomal structure nor did they appear in monolayer cultures. They might be considered as the sites for ion exchange. The nerve-fiber network was composed at this stage of development of bundles of naked fibers of different diameters (Fig. 7b). During the following two weeks in culture, cells reached maturation, synaptogenesis and myelination. Perikarya became larger in size and in their cytoplasm, a well developed Nissle substance, Golgi cisternae and many mitochondria could be recognized. In the neuropil, there were many synaptic boutons and axodendritic synapses along with astrocytic fibers containing numerous microfilaments. Most of the synapses were mature and contained clear synaptic vesicles of the cholinergic type (Fig. 8a, b). Often, single electrondense core vesicles were also present. Acetylcholinesterase (AChE) activity was measured according to Johnson and Russel [54], in monolayer and MC cultures at various times during three weeks of cultivation. MC cultures showed a 4–5 fold increase in enzyme activity when cultured in both ways: stationary MC cultures (in 32 mm coated plastic dishes) or suspended MC cultures (in 100 ml flasks, stirred at 40 rounds per minute). In TEM a higher number of synapses was counted in sections from MC cultures in comparison with monolayer cultures.

Myelinated axons became apparent after the second week in culture [6, 29]. Myelination increased progressively and in about a month, in vitro cultures appeared heavily myelinated. A compact multilamellar myelin sheath, characteristic of the CNS, enwrapped most of the axons. Several fibers however, as well as single perikarya appeared contornated by myelin sheath composed of only a few lamellae (Fig. 9a, b, c).

5 Primary Cultures of Myoblasts

5.1 Monolayer Cultures

The conventional way of growing embryonic myoblasts in vitro is in monolayer primary cultures [55, 56]. Dissociated mononucleated myoblasts undergo morphological and biochemical differentiation in these cultures. They fuse into multinucleated myotubes which become striated and contract spontaneously intensively. During myogenesis in culture, two types of proteins are formed by the cells: (1) A cytoplasmatic protein, such as creatine-kinase known to be an indicator of muscle differentiation [57]; (2) Membrane proteins like acetylcholin receptors and acetylcholinesterase which are known to be influenced by muscle contractions and its state of innervation [58, 59, 60]. The levels of these proteins serve as a probe for the study of stages in muscle development and regeneration. Although the stages of myoblast differentiation in vitro are similar to those of striated muscle in vivo, myotubes in monolayer cultures are oriented randomly and are not packed parallel to each other to mimic the in vivo

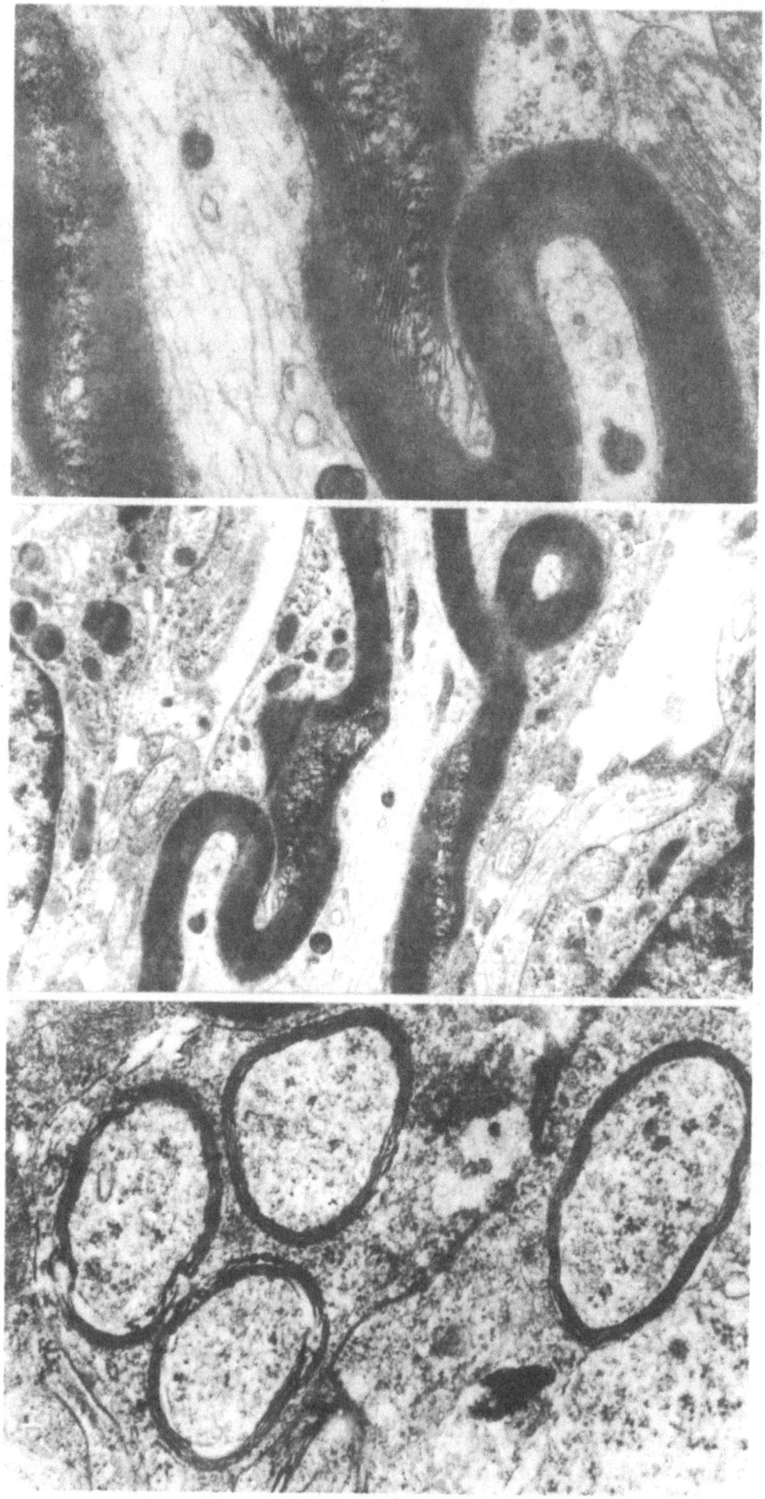

a

b

c

spacial orientation. The main limitations of the monolayer culture technique are: Peeling off the dish of muscle fibers probably due to active contractions (usually after 2–3 weeks of cultivation) and the difficulty in collecting samples from ongoing cultures.

5.2 Cultures on MCs

In order to overcome these disadvantages, we have grown myoblasts from chick embryo, rat and mouse fetuses on cylindrical DE-53 DEAE cellulose MCs in stationary cultures [24, 30, 31].

The growth pattern of dissociated myoblasts on MCs was basically similar to that of monolayer cultures. However, the choice of cylindrical, rather than spherical MCs as used previously by others [25], was mainly to provide a larger support for the development of the numerous elongated myotubes. Most of the cells attached to the MCs within the first 24 h to form cell — MC conglomerates. These conglomerates are much larger than those formed in CNS dissociated MC cultures (Fig. 10a). They remain floating for the entire period of cultivation and contract spontaneously. Protein synthesis, as well as RNA and DNA replication, although at a lower rate than in monolayer cultures, proceeded progressively, and by the end of the first week, the fusion of myoblasts into myotubes, was almost completed.

In SEM, bundles of flat spindle shaped fibers were attached to the MCs parallel to each other, in the same spacial orientation. A few cells were in mitosis (Fig. 11b). The fibers became thicker during the second week in culture. They exhibited on their surface spherical microprojections and emanated on their lateral sides equally spaced elongated tangled microvilli (Figs. 11b, 12a). Large swellings, indicative of nuclear location were noticeable in the fiber's cytoplasm (Fig. 12a). Within the conglomerates, several MCs formed an anchorage for the fiber endings (Fig. 12b). In TEM fibers showed the ultra structural organization of a developed sarcomer (Fig. 10b). Myofibrils were organized in bundles revealing characteristic cross striations. Among other cytoplasmatic organells, there were many mitochondria, in part aligned along the myofibrils; several cisternae of Golgi apparatus, sarcoplasmic reticulum, and a typical smooth endoplasmic reticulum with a honeycomb-like structure. Myotubes exhibited vesicular nuclei with a few lobulations and prominent nucleoli.

Biochemical differentiation of myoblasts on MCs was similar to that achieved in monolayer cultures. Measurements of levels of both creatine kinase and acetylcholine receptors were used as markers. A concomitant increase in the synthesis of these two proteins occurred during fusion of myoblasts to reach a plateau after two weeks in culture. Levels were slightly higher in MC cultures than in monolayers (Fig. 13). In contrast to monolayer cultures which peeled off the dish toward the 3rd. week of cultivation, the synthesis of these two proteins could be measured in MC cultures for a period of three months in vitro [24, 30].

◀

Fig. 9a–c. Axons surrounded by a multilamellae myelin sheath. **a** and **b** and an unidentified structure contornated by a few lamellae of myelin, **c**. **a**, ×40000; **b**, ×20000 and **c** — ×40000

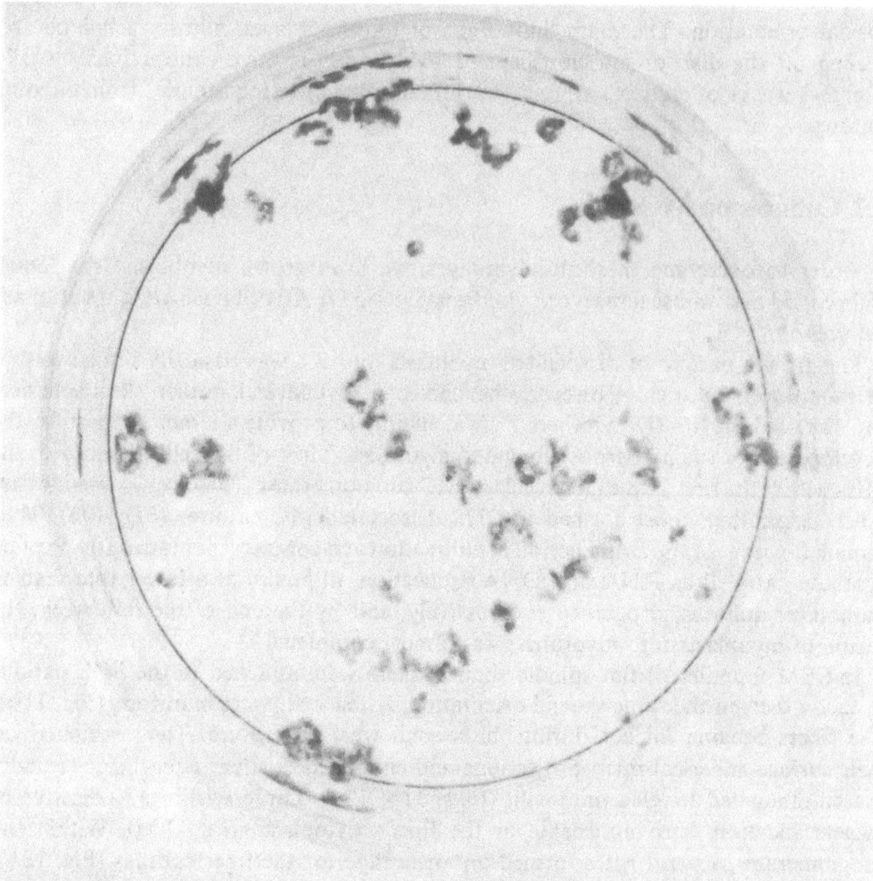

a

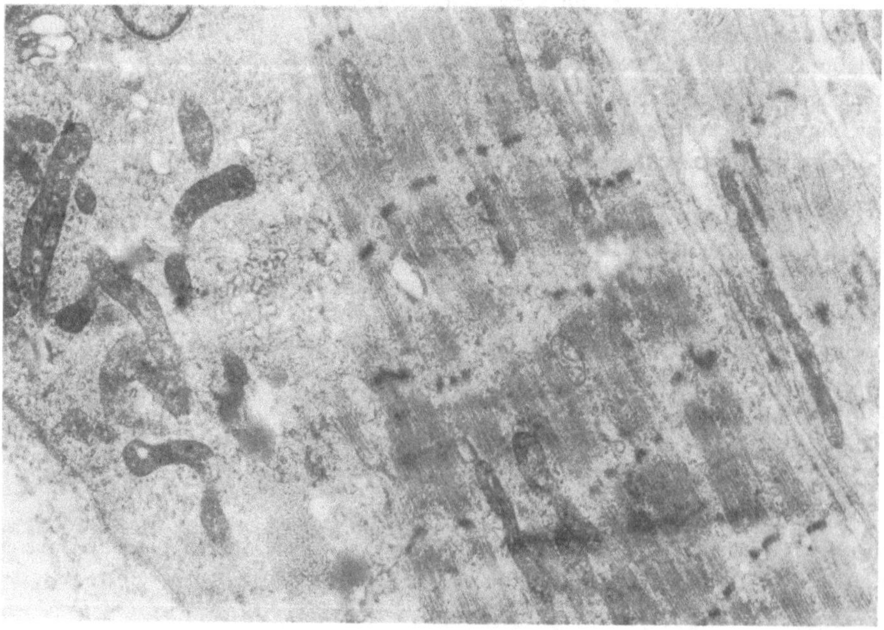

b

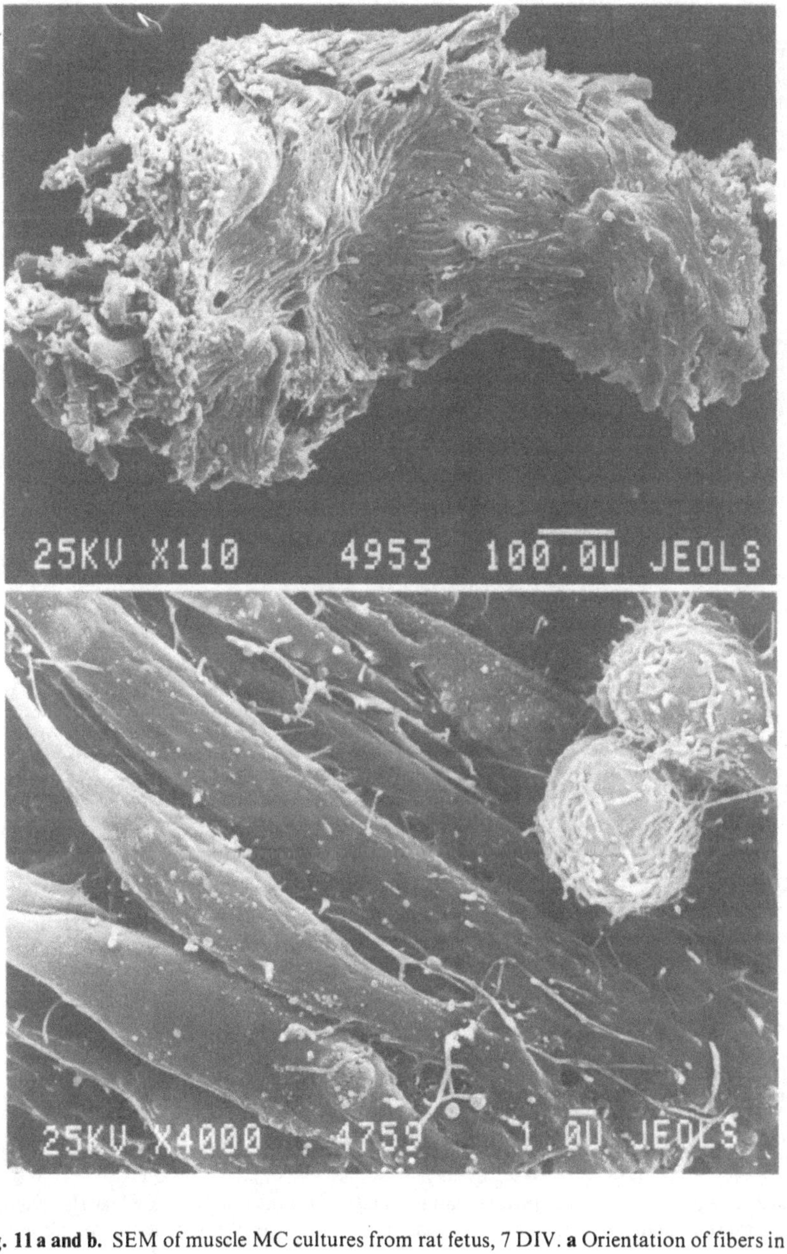

a

b

Fig. 11 a and b. SEM of muscle MC cultures from rat fetus, 7 DIV. **a** Orientation of fibers in conglomerate; **b** Parallel myotubes and a cell in mitosis

◀

Fig. 10 a. Muscle MC conglomerates from 12 d chick embryo, 5 DIV, ×3.5; **b**, TEM of sarcomeric organization in MC cultures of chick embryo, 14 DIV. Note: At the bottom left, a honeycomb-like agranular endoplasmic reticulum. ×10800

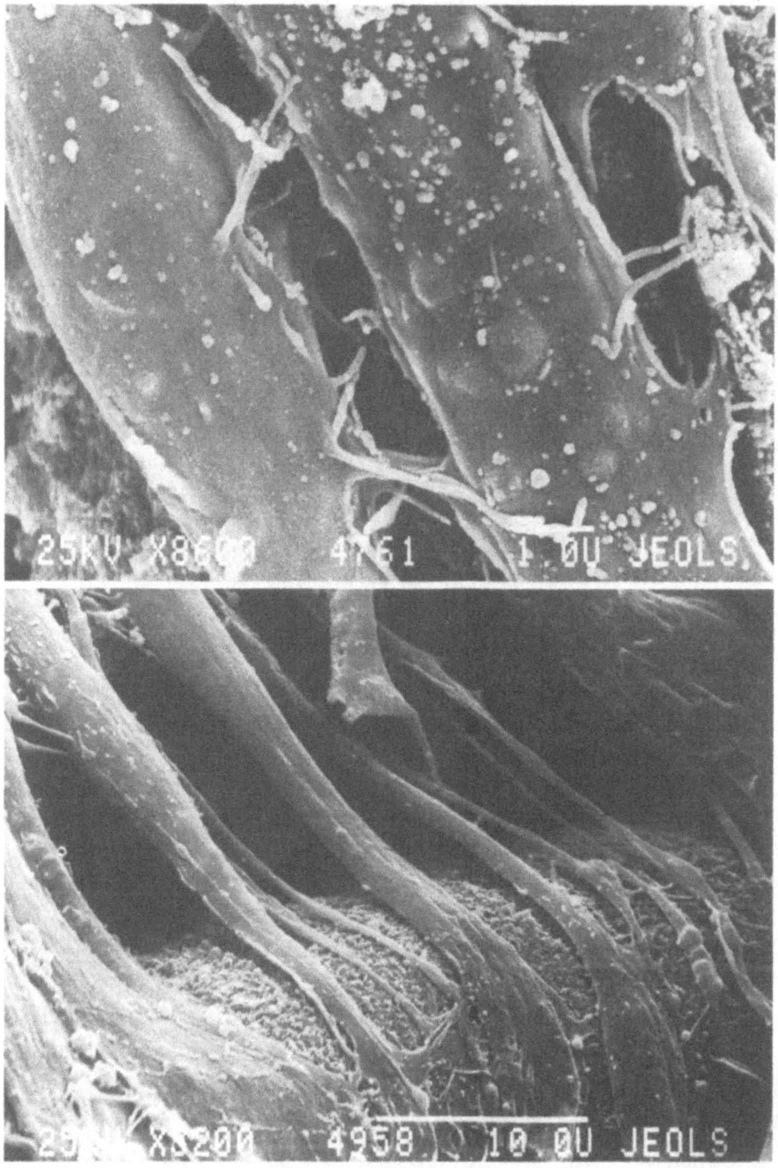

a

b

Fig. 12a and b. Myotubes developed in a MC culture, 7 DIV, showing in **a**, Elongated tangled microvilli on their lateral sides and swellings of nuclear areas; **b** Terminations of myotubes surrounding a MC

6 Conclusions and Prospective Remarks

The major achievement of the MC primary cultures is in providing optimal conditions for embryonic dissociated CNS cells and myoblasts to organize and differentiate into functional neuronal and muscular units. The neuronal entities consist of newly

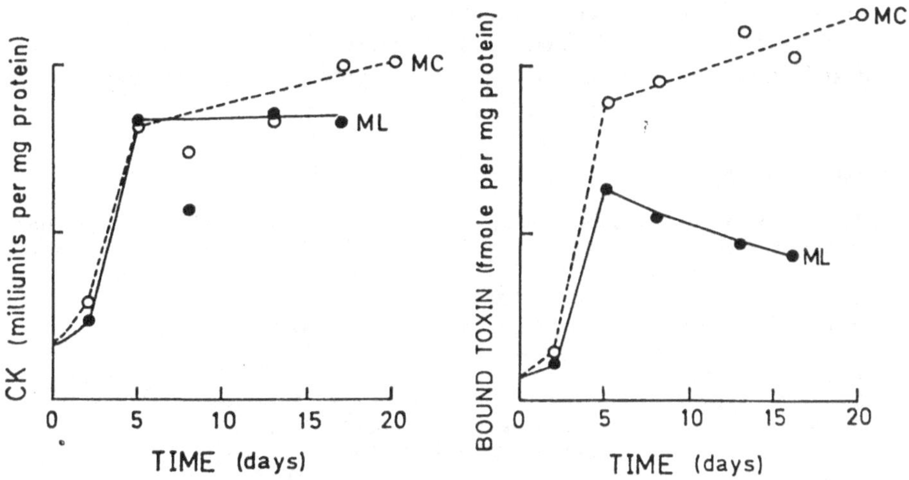

Fig. 13. Comparison of measurements of creatine kinase and acetylcholin receptors in monolayer and MC muscle cultures. Each point is an average of 2–3 replicates

developed neuropils with numerous synapses and a well established nerve fiber net work with many myelinated axons. The muscular units are composed of groups of striated myotubes which are oriented in the same direction and contract spontaneously. Such entities remain floating in the culture medium and can be harvested for morphological and biochemical analysis at any given moment without interfering with the ongoing culture. This technique is advantageous over the conventional monolayer culture procedure, since it provides a tridimensional anchorage for neuronal and muscle cells which in this way can be grown to a higher amount at a better efficiency and for longer periods (months).

Neuronal and muscle cultures have recently been recognized as a valid in vitro tool for toxicological evaluation of chemicals, drugs, environmental toxins, etc. Maturation, long-term culturing and easy sampling of these cells when grown in MC-cultures, makes this culture system a model of choice for analysing the effects of long-term exposure to controlled concentrations of single and multiple toxicological agents. One possible use of cultured neuronal and muscle — MC entities, is their implantation into injured or dystrophic tissues in order to investigate the role played by the transplanted embryonic material in adult tissue regeneration. In preliminary experiments (manuscript in preparation), embryonic myogenic cells which had been cultivated on MCs and labelled with ^{3}H-thymidine (1 μC per ml) were implanted into the injured area of the gastrocnemius muscle of adult rats. The autoradiographical studies suggested active involving of the transplanted embryonic material in the regeneration process. The presence of the transplanted material prevented on the one hand, the formation of a connective tissue scar and on the other hand accelerated the formation of new myotubes, located around the MCs. These observations indicate that the implanted myoblasts grown on MCs in vitro, might participate (with satellite cells) in the process of adult muscle regeneration.

Finally co-cultivation of neurons and muscle cells on MCs has the possibility to become a useful model for the study of the reciprocal effects on interaction, organization and differentiation of these two tissues when grown on a tridimensional support.

7 References

1. Reuveny, A.: Microcarriers for culturing mammalian cells and their application, in: Advances in Biotechnological Processes (eds. Mizrahi, A., Van Wezel, A. L.) Vol. 2, p. 1, New York, Alan R. Liss 1983
2. Davis, P. F.: Exp. Cell Res. *134*, 367 (1981)
3. Ryan, U. S., Mortara, M., Whitaker, C.: Tissue Cell *12*, 619 (1980)
4. Horst, J., Kern, M., Ulmer, E.: Eur. J. Cell Biol. *22*, 599 (1980)
5. Duda, E.: Cryo. Lett. *3*, 67 (1982)
6. Shahar, A., Reuveny, S., Amir, A., Kotler, M., Mizrahi, A.: J. Neurosci. Res. *9*, 339 (1983)
7. Sargent, G. F., Sims, T. A., McNeish, A. S.: J. Microsc. *122*, 209 (1981)
8. Crespi, C. L., Thilly, W. G.: Mutation Res. *106*, 123 (1982)
9. Wright, J. T., Elmer, W. A., Dunlop, A. T.: Anal. Lab. *125*, 100 (1982)
10. Lai, C. S., Hopwood, L. E., Swartz, H. H.: Exp. Cell Res. *130*, 437 (1980)
11. Hoeg, J. M., Osborne, J. C., Brewer, H. B.: J. Biol. Chem. *257*, 2125 (1982)
12. Vosbeck, K., Roth, S.: J. Cell Sci. *22*, 657 (1976)
13. Fairman, K., Jacobson, B. S.: Tissue Cell *15*, 167 (1983)
14. Kotler, M., Reuveny, S., Mizrahi, A., Shahar, A.: Develop. Biol. Stand. *60*, 255 (1985)
15. Davis, P. F., Kerr, C.: Exp. Cell Res. *141*, 455 (1982)
16. Davis, P. F., Ganz, P., Diehl, P. S.: Lab. Invest. *85*, 710 (1985)
17. Ren, E. C.: J. Immunol. Meth. *49*, 105 (1982)
18. Howard, G. A., Turner, R. T., Fuzas, J. E., Nichols, F., Baylink, D. J.: JAMA *249*, 258 (1983)
19. Fayet, B., Hovsepian, S.: Biochimie *61*, 923 (1979)
20. Smith, M. A., Vale, W. W.: Endocrinology *107*, 1425 (1980)
21. Smith, M. A., Vale, W. W.: ibid. *108*, 752 (1981)
22. Spiess, Y., Smith, M. A., Vale, W.: Diabetes *31*, 189 (1982)
23. Busch, C., Cancilia, P. A., DeBault, L. E., Goldsmith, J. C., Owen, W. G.: Lab. Invest. *47*, 498 (1982)
24. Shainberg, A., Isac, A., Reuveny, S., Mizrahi, A., Shahar, A.: Cell Biol. Int. Rep. *7*, 727 (1983)
25. Pawlowski, R., Szigeti, V., Loyd, R., Pryzbylski, R. J.: Eur. J. Cell Biol. *32*, 296 (1983)
26. Ostlund, C., Clark, J., Kruse, M.: In Vitro *19*, 279 (1983)
27. Bone, A. J., Swenne, I.: ibid. *18*, 141 (1982)
28. Reuveny, S., Mizrahi, A., Shahar, A., Kotler, M.: Cell Biol. Int. Rep. *8*, 7, 539 (1984)
29. Shahar, A., Amir, A., Reuveny, S., Silberstein, L., Mizrahi, A.: Develop. Biol. Standard *55*, 25 (1984)
30. Shahar, A., Mizrahi, A., Reuveny, S., Zinman, T., Shainberg, A.: ibid. *60*, 263 (1985)
31. Shahar, A., Reuveny, S., Mizrahi, A., Shainberg, A.: J. Acad. Med. Torino *CXLVII*, 33 (1984)
32. Reuveny, S., Silberstein, L., Shahar, A., Freeman, E., Mizrahi, A.: In Vitro *18*, 92 (1982a)
33. Reuveny, S., Silberstein, L., Shahar, A., Freeman, E., Mizrahi, A.: Develop. Biol. Stand. *50*, 115 (1982b)
34. Hirtenstein, M., Clark, J., Lindgren, G., Vertblad, P.: ibid. *46*, 109 (1980)
35. Lewis, D. H., Volkers, S. A. S.: ibid. *42*, 147 (1979)
36. Gebb, C., Clark, J. M., Hirtenstein, M. D., Lindgren, G., Lindgren, U., Lindskog, U., Lundgren, B., Vertblad, P.: ibid. *50*, 93 (1982)
37. Paris, M. S., Eaton, D. L., Sempolinski, D. E., Sharma, B. P.: In vitro *19*, 262 (1983)
38. Johansson, A., Nielsen, V.: Develop. Biol. Stand. *46*, 125 (1980)
39. Nielsen, V., Johansson, A.: ibid. *46*, 131 (1980)
40. Varani, J., Dame, M., Beals, T. F., Wass, J. A.: Biotech. Bioeng. *25*, 1359 (1983)
41. Varani, J., Dame, M., Rediske, J., Beals, T. F., Hellegas, W.: J. Biol. Stand. *13*, 67 (1985)
42. Gahwiler, B. H.: Neuroscience *11*, 751 (1984)
43. Sobkowicz, A. M., Bleier, R., Bereman, B.: J. Neurocytol. *3*, 431 (1974)

44. Gahwiler, B. H.: Experientia *40*, 235 (1984)
45. Yavin, Z., Yavin, E.: Exp. Brain Res. *29*, 137 (1977)
46. Shahar, A. et al.: Aging in cerebral and motor neurons of fetal and adult origin, in: The Aging Brain: Cellular and Molecular Mechanisms of Aging in the Nervous System (eds. Giacobini, E. et al.) p. 35, New York, Raven Press 1982
47. Sensenbrenner, M.: Proliferation and maturation of neuronal cells from the central nervous system in culture, in: Neurotransmitter Interaction and Compartmentation (ed. Bradford, H. F.) p. 497, Plenum Publishing Corporation 1982
48. Sensenbrenner, M.: Dissociated brain cells in primary cultures, in: Cell, Tissue and Organ Cultures in Neurobiology (eds. Fedoroff, S., Hertz, L.) p. 191 New York, Academic Press 1977
49. Seeds, N. W., Ramirez, G., Marko, M. J.: Aggregate cultures: A model for studies of brain development, in: Cell Culture and its Application (eds. Acton, R. T., Lynn, J. D.), New York, Academic Press 1977
50. Honegger, P.: Brain Res. *162*, 89 (1979)
51. Matthieu, J. M., Honegger, P., Favrod, P., Poduslo, J. F., Ceccarini, C., Kristic, R.: Myelination and demyelination in aggregating cultures of rat brain cells, in: Tissue Culture in Neurobiology (eds. Giacobini, E., Vernadakis, A., Shahar, A.), New York, Raven Press 1980
52. Matthieu, J. M., Honegger, P., Trapp, B. D., Cohen, S. R., Webster, H. De.: Neuroscience *3*, 565 (1978)
53. Seeds, N. W., Haffke, S. C.: Dev. Neurosci. *1*, 69 (1978)
54. Johnson, C. D., Russel, R. L.: Anal. Biochem. *64*, 229 (1975)
55. Yaffe, D.: Current Topics in Develop. Biol. *4*, 37 (1969)
56. Shainberg, A., Yagil, G., Yaffe, D.: Develop. Biol. *25*, 1 (1971)
57. Shainberg, A.: Nature *264*, 368 (1976)
58. Fambrough, D. M.: Physiol. Rev. *59*, 165 (1979)
59. Koenig, J.: Prog. Brain Res. *49*, 484 (1979)
60. Shainberg, A., Shahar, A., Burstein, M., Giacobini, E.: Effect of innervation on acetylcholine receptors in muscle cultures, in: Tissue Culture in Neurobiology (eds. Giacobini, E. et al.) p. 25, New York, Raven Press 1980

Growth Limitations in Microcarrier Cultures

M. Butler
Department of Biological Sciences, Manchester Polytechnic, Chester Street, Manchester MI 5GD, U.K.

The use of various commercially avaiable microcarriers is considered for the large-scale growth of anchorage-dependent animal cells. Cell productivity in such cultures can be maximised by an analysis of various kinetic aspects of growth. Such parameters as the critical cell to bead ratio at inoculation is particularly important in this respect.

The full theoretical capacity of microcarriers to support cell growth can only be realised by an understanding of the interaction of the cell with is micro-environment. The final cell yield in culture may be limited by the supply of nutrients or by the accumulation of inhibitory metabolites. Glucose and glutamine are widely used as the major energy sources for cell growth and their relative concentrations in the media can influence cellular metabolism. The products of such metabolism — lactic acid and ammonia — may be inhibitory and methods to reduce their accumulation should be considered. The requirements for other essential metabolites such as amino acids vary but their supply should reflect the needs of each specific cell type. The advantages of serum-free media and the problems of oxygen supply particularly in large-scale microcarrier cultures are reviewed.

Full investigation of these parameters that affect cell productivity will result in an increased ability to control the culture of anchorage-dependent animal cells on microcarriers. This will lead to considerable benefits for the production of an expanding list of useful cell products.

1 Introduction

The development of processes for the large scale culture of animal cells was largely precipitated by the need for mass production of viral vaccines in the 1950s. The animal cells considered suitable for vaccine production were normal, diploid and anchorage-dependent. Because such cells closely resemble those in the natural physiological state, they are regarded as safe substrates for the production of biologicals intended for medical and veterinary use. Since the 1950s the need for production processes for such anchorage-dependent cells has increased with the expanding list of their useful products.

The major differences in the design of a process for the growth of these cells compared to the traditional microbial culture systems is that a solid substratum is required for growth. At the laboratory scale the substratum can be provided by the flat surfaces offered by Petri dishes, T-flasks or Roux bottles. However, these are not amenable to scale-up because of the larger growth surface areas required. This problem was originally solved by the use of roller bottles. In such a system the constant movement of the culture medium around the inner surface of each rolled cylindrical bottle allows the full use of the inner surface for the growth of a cell monolayer. A viral vaccine production process based on this system may require several thousand roller bottles. There are a number of disadvantages of such a multiple culture process. Each roller bottle must be handled individually and so a simple media change may require several thousand sterile manipulations. Thus the process is labour intensive and subject to excessive risk of contamination. A further disadvantage is the difficulty of culture control when so many independent bioreactors are involved.

For these reasons considerable attention has been paid to the possibilities of replacing such a multiple reactor system with a unit reactor of high cell productivity. The objective in designing a suitable unit bioreactor is the provision of a large surface area per unit volume. Simple modifications of the standard roller bottle system by inclusion of stacked multi-plates or spiral plastic films can lead to some increase in the surface to volume ratio. However, to be able to obtain much higher ratios, completely different types of reactors are required. These include glass bead propagators, artificial capillary systems, tubular spiral systems and microcarrier suspensions.

A comparison of the surface to volume ratios offered by various bioreactors (Table 1)

Table 1. Surface-to-volume ratio (cm^{-1}) offered by various methods for the culture of anchorage-dependent animal cells [1]

Culture system	S/V ratio
Roller bottle	1.25
Gyrogen	1.2
Multi-plate bottle	1.7
Spiral film bottle	4.0
Plastics bags	5.0
Tubular spiral film	9.4
Packed bed reactor	10.0
Artificial capillaries	30.7
Microcarriers (25 $g\,l^{-1}$)	150.0

shows that the microcarrier suspension system offers the best potential for high productivity of anchorage-dependent cells. Further attractive features of the microcarrier system include the ease of culture sampling and the ability to use the stirred tank fermenters which were originally designed for microbial cultures. Because of these features, considerable attention is being paid to the use of microcarriers for animal cell growth particularly for obtaining high cell density cultures.

2 Development of Microcarriers

The concept of using a suspension of microspheres to allow a high surface-to-volume ratio for the growth of anchorage-dependent cells in a single bioreactor was developed by van Wezel [2]. As well as providing a large surface area for cell growth, it was recognised that the system could be made homogeneous by constant stirring. This provides a quasi-suspension culture and has several advantages:

a: Sampling can be made relatively easy

b: The danger of nutrient gradients within the culture is reduced. Thus the microenvironment of each cell can be kept the same.

c: The cellular environment can be controlled by whole media changes or selective additions.

The microcarrier originally chosen by van Wezel [2] was DEAE-Sephadex A-50 which was designed primarily for ion exchange chromatography. These microcarrier beads consist of a dextran matrix to which is bound, diethylaminoethyl groups which provide a net positive charge considered necessary for cell attachment. The microcarriers were found suitable for cell growth at a particle concentration of 1 g l^{-1}. However at microcarrier concentrations above this level, adverse effects were observed on the cells which tended to detach from the surface. This was considered to be related to a "toxic" effect emanating from the beads.

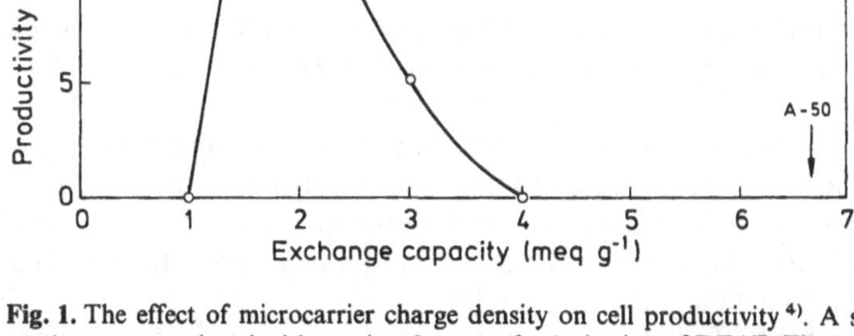

Fig. 1. The effect of microcarrier charge density on cell productivity [4]. A series of dextran microcarriers were produced with varying degrees of substitution of DEAE. The cell productivities (10^6 per L h^{-1}) of human diploid fibroblasts (HEL 299) were determined in microcarrier cultures at 5 g dextran per L. The arrow indicates the exchange capacity of DEAE-Sephadex A-50

The "toxic" effect of these beads could be reduced by allowing a protein to coat the charged surface prior to cell attachment. However, it became apparent that this "toxicity" was related to the charge density on the bead surface. A reduction of the positive charge of DEAE Sephadex A-50 from 6.5 to 1.8–2.0 meq g^{-1} allowed cell attachment and growth at higher microcarrier concentrations [3] (Fig. 1). Such reduced charge DEAE dextran microcarriers have been found suitable for the growth of a large variety of cell types and have been commercially available for a number of years as — Superbeads (Flow Labs) and Cytodex (Pharmacia).

3 Characteristics of Microcarriers

There are many features and properties that may be considered in the design of an ideal microcarrier for cell growth. These include:
a) Density. Animal cells may suffer damage at high shear forces and to minimize such damage cultures must be maintained at low stirring speeds. The optimum microcarrier density to maintain an even suspension in culture is 1.03 g ml^{-1}. Beads of higher density settle to the bottom of the bioreactor whereas lighter beads tend to float — both of these characteristics being undesirable. The density of each bead does increase slightly as cell attachment and growth occurs. Thus it may be necessary to increase the stirring speed after initial attachment [5]. This increased density may be useful in microcarrier sedimentation during medium perfusion [6,7].
b) Diameter. In order to maximise the growth surface per unit volume of the bioreactor, a high concentration of small beads is required. However, each bead must have sufficient surface area to support the growth of a single viable cell over several generations. A bead diameter of 100 to 400 μm has been found to be suitable. Each bead of this size will support about 10^2 to 10^3 cells as a monolayer.
c) Size distribution. The size distribution of the microcarriers needs to be small in order to ensure an even cell distribution on beads at inoculatation. An uneven size distribution of microcarriers results in selective cell attachment to the smaller beads [8].
d) Charge. This parameter caused the greatest problem in the development of microcarriers. The surface charge density of each dextran bead must be optimised at 2.0 $\pm$ 0.5 meq g^{-1} before cell growth can be supported. Figure 1 shows that very little deviation can be tolerated from this optimal charge which is suitable for the growth of many cell lines [3].
e) Transparency. The transparency of microcarriers is important during microscopic observation. Stained cells attached to microcarriers can then be clearly observed under a microscope.
f) Porosity. This should be low so that the concentration of added proteins such as growth factors and serum is not reduced in culture by adsorption to the microcarriers.
g) Rigidity. Damage to microcarriers or cells may result from collisions in culture. To minimise such effects, microcarriers should be sufficiently strong to withstand collision but non-rigid to minimise cell damage.
h) Surface binding material. The strength and nature of cell binding depends upon the surface material of the microcarrier. This is particularly important in considering the attachment of cells at inoculation and detachment during harvesting.

4 Types of Microcarriers

The original microcarriers that were developed were modifications of the chromatographic grade DEAE Sephadex A-50. These are dextran beads to which the positively charged tertiary amine — N,N diethylaminoethyl — is cross linked at a specified concentration. This formulation has been found suitable for the growth of numerous cell lines and types [5,9] (Fig. 2) These dextran beads are available as Superbeads and Cytodex 1 and are probably the most widely used microcarriers.

There are two major parameters of microcarrier design that may be altered from the original DEAE-dextran type whilst still maintaining acceptability for cell growth.

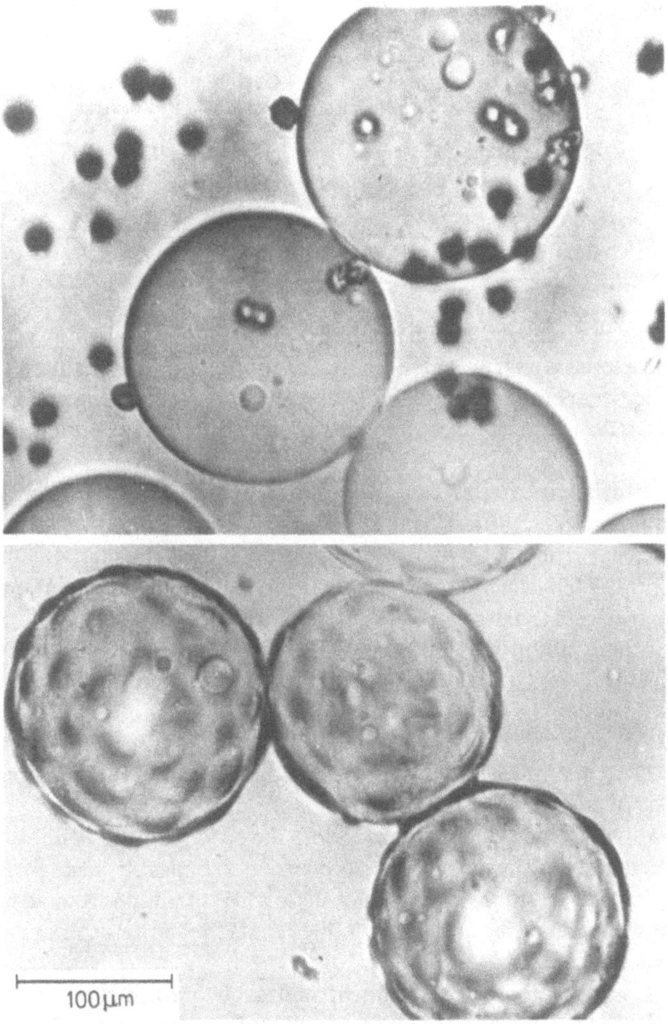

a

b

Fig. 2 a. Canine epithelial cells (MDCK) after inoculation into a DEAE-dextran microcarrier suspension; **b.** Confluent monolayers of MDCK cells on the surface of DEAE-dextran microcarriers

Table 2. Microcarriers which are commercially available

Type	Trade mark	Company	Country	Composition
a) Dextran				
	Superbeads	Flow Labs	USA	DEAE-dextran
	Cytodex 1	Pharmacia	Sweden	DEAE-dextran
	Cytodex 2	Pharmacia	Sweden	Quaternary amine-coated dextran
	Microdex	Dextran Products	Canada	DEAE-dextran
b) Plastic				
	Biosilon	Nunc	Denmark	Polystyrene — charged
	Biocarriers	Biorad	USA	Polyacrylamide/DMAP
	Cytobeads	Lux	USA	Polystyrene — charged
	Acrobeads	Galil	Israel	Polyacrolein — various coatings
d) Gelatin				
	Geli-beads	KC Biologicals	USA	Gelatin
	Ventregel	Ventrex	USA	Gelatin
	Cytodex 3	Pharmacia	Sweden	Gelatin-coated dextran
e) Glass				
	Bioglas	Solohill Eng.	USA	Glass-coated plastic/latex
f) Cellulose				
	DE-52/53	Whatman	UK	DEAE-cellulose

a) The basic matrix of the microcarrier may be altered.

b) The surface coating material which provides the electrostatic charge may be altered.

Microcarriers with various characteristics with respect to these parameters are produced commercially and a listing of those which have come to the author's attention is shown in Table 2.

Two alternative dextran microcarriers have been developed by Pharmacia, with different surface coat materials — Cytodex 2 and Cytodex 3. Cytodex 2 has a quaternary amine group — N,N,N-trimethyl-2-hydroxyaminopropyl, distributed only on the surface of the bead [5, 10]. The surface charge density for cell attachment is maintained wheras the total charge of the bead is reduced. Consequently the adsorption of any ionic groups from the medium is minimized. This may be advantageous if the adsorption of growth factors or cell products which may be at low concentrations, were to become a problem. Cytodex 3 consists of dextran beads which are coated with denatured collagen (i.e. gelatin) [5, 10]. This gelatin coat allows the easy attachment of cells which may have a low plating efficiency. A further advantage of such a coating material may be the ease of harvesting cells in a state of high viability.

The use of gelatin as a support material has been made in the production of Gelibeads (KC Biologicals) and Ventregel (Ventrex). These microcarriers are entirely made from gelatin and cells may be harvested by completely dissolving the support matrix. Trypsin, collagenase or dispase have been found suitable agents for cell detachment — either individually or as a mixed cocktail [11].

Glass was used as a support for cell attachment and growth from the earliest development of small scale laboratory techniques of cell culture. Its use can be extended to a large-scale in packed bed bioreactors [12, 13] in which 2—3 mm diameter solid glass beads are used. However, glass is not a suitable material as a basic matrix for microcarriers because of its high density — 1.5 g cm^{-3}. Thus adequate suspension in liquid medium could not be maintained at moderate stirring rates. However a microcarrier consisting of a plastic matrix and a coating of glass is now commercially available — Bioglas (Solohill Eng. Inc.) — and this has an acceptable density. Glass is an efficient surface for cell attachment and removal by proteolytic enzyme treatment. Furthermore, these microcarriers may be easily cleaned and re-used [14)].

Cellulose has been used successfully as a microcarrier matrix in the form of the microgranular DEAE-cellulose anion exchange material — Whatman DE-52 and DE-53 [15,16,17)]. DE-53 has a positive charge capacity similar to the DEAE-dextran beads — Cytodex 1, Superbeads or Microdex — and supports the growth of a number of cell types to an acceptable density [18)]. DE-52 is more limited in the cells whose growth can be supported because of its lower charge capacity. The novel characteristic of these microcarriers is the elongated cylindrical shape.

Plastic is used as a matrix for many commercially produced microcarriers. The main advantages of using any plastic material are that the microcarriers are shape stable, non-swelling and relatively resistant to breakage. Biosilon (Nunc), Cytobeads (Lux) and Bio-Beads (Biorad) are produced from treated polystyrene which has been found to be suitable for the production of much of the disposable plastic products for laboratory cell culture [19)]. The polystyrene requires treatment usually by sulphonation to an optimal negative surface charge [20)] before it is suitable for cell attachment and growth [21)]. Glycine-derivatized polystyrene has been found optimal for several cell lines [(22)]. Although optimal cell growth can be maintained on these polystyrene beads, some cases have been reported where the cells detach too readily after the beads become confluent [23)]. Another feature worthy of consideration is that although microscopic examination of these beads is possible, they are not as translucent as dextran based beads [22)] and this may result in difficulties in observing growing cells [24)].

Other plastic materials that have been used for microcarrier production include polyacrylamide [25,26,27,28)], trisacryl [29)] and polyacrolein (Acrobeads from Galil). These materials require the addition of charged groups or coating materials before they are suitable for supporting cell growth. Tertiary amino groups such as DEAE can be used successfully but for selected cell lines, preferential attachment may occur on beads charged with primary amino groups [25)] which may be attached to hydrophobic hydrocarbon chains [27)].

The use of fluorocarbon emulsions coated with polylysine has been found suitable for microcarrier cell growth on a small scale [30)]. The droplet size distribution is large (100—200 µm) but the system has the advantage of simple cell harvesting by centrifugation which breaks the emulsion into its component phases.

All the microcarriers mentioned above conform to the essential requirements considered earlier and have been reported in the literature as supporting cell growth. Differences in performance arise through preferential attachment of sensitive cell lines on certain microcarrier formulations. In this respect it is interesting to note that although the surface charge density of the DEAE-dextran microcarriers was found to be particularly important [3)] for cell attachment, the polarity of this charge may be altered. Thus many of the plastic and glass based microcarriers may be negatively charged e.g. sulphonated polystyrene, glass coated beads. The cell membrane surface carries a net negative charge [31,32)] which may be unevenly distributed [32)] and depend upon the physiological state of the cell [33)]. However the mediators of cell adhesion to a surface are divalent cations and glycoproteins. The latter can be provided by serum components or by secretion from the cells. It is the amphoteric nature of the binding glycoproteins in the presence of an ionic environment that allows cell attachment to electrostatic surfaces of either polarity [34].

The nature of the cell-substratum attachment on microcarriers was studied by Burke et al. [35] and Fairman and Jacobson [36]. They showed a unique morphology of attached cells when the microcarrier is coated with a protein such as collagen for which the cell has specific receptors [36]. This suggests a distinct mode of attachment to that observed on non-specific substrata [35]. Their analysis of cell attachment on polystyrene beads under high shear showed that the rates of attachment were independent of the polarity of the charged beads or whether the beads were coated with gelatin or BSA. Only when beads were coated with laminin or fibronectin was a decreased rate of attachment shown.

Certain microcarriers have been developed to allow easy cell detachment and harvesting. However the original difficulties found in the proteolytic removal of cells from DEAE-dextran beads [37] can be reduced by the use of a recently developed protocol for high pH trypsin removal [38]. This is effective in harvesting high viability human diploid fibroblasts from microcarriers.

The choice of a particular microcarrier should depend upon:

a) The ability of the selected cell line to attach to the microcarrier surface.

b) The prefered method of cell harvesting from the microcarrier surface. For most purposes the correctly charged DEAE-dextran microcarrier is adequate. However, all the microcarriers discussed offer a large surface area-to-volume ratio for growth and are suitable for large scale cultures. At the present state of the art, surface area is not likely to be the factor in limiting cell productivity. For most cell lines a density of 2–3 × 10^6 cells per ml can be achieved in a culture containing any one of the number of commercially available microcarriers.

5 Kinetic Aspects of Cell Growth on Microcarriers

5.1 Theoretical Capacity of Microcarriers

The potential of microcarriers for supporting cell growth can be realized from a theroretical calculation of the available surface area compared to one of the traditional systems e.g: a Roux bottle.

a) *Roux bottles*
   Surface area available for growth per bottle = 2×10^2 cm^2
   Cell number typically obtained per bottle = 10^8

b) *Microcarriers*
   Mean diameter = 170 μm
   Mean surface area per microcarrier = 10^{-3} cm^2
   Number of microcarriers per mg dry weight = 6,300
   If microcarriers are suspended in a 1 litre culture at 2 mg ml^{-1}:—
   Surface area available for growth = 10^4 cm^2 L^{-1}
   Cell yield typically obtained = 2×10^9 L^{-1}

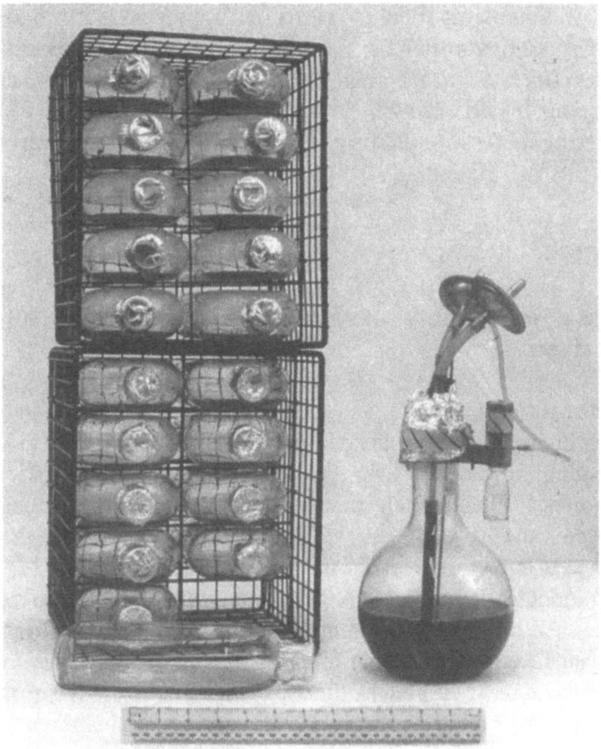

Fig. 3. A 1 L culture of microcarriers compared to a set of 20 Roux bottles. Both have the equivalent potential for cell productivity

c) *Comparison of microcarrier capacity vs roller bottles*
On a surface area basis:
1 L microcarrier culture = 50 Roux bottles
On a cell yield basis:
1 L microcarrier culture = 20 Roux bottles

This latter comparison illustrates the major attraction of the microcarrier system. The handling operations involved in a 1 litre microcarrier culture are considerably simpler than involved in 20 Roux bottle cultures and an equivalent cell yield can be obtained (Fig. 3). A typical microcarrier culture involves an initial cell inoculation of 2×10^5 cells per ml which leads to a final cell density of $\sim 2 \times 10^6$ cells per ml. Thus a cell multiplicity of $\times 10$ is obtained in the culture. In order to obtain these cell yields concentrations of 1 to 3 mg microcarriers per ml are typically used.

At this level of cell productivity the microcarrier system has clear advantages for scale-up. However, the full potential of microcarrier cultures may not have been realised at this level of cell productivity. The surface area available for cell growth in a simple batch culture may be increased by raising the concentration of microcarriers. Thus by linear extrapolation from these calculations a microcarrier suspension of 10–20 mg ml^{-1} should realise a cell yield of 10×10^6–20×10^6 cells per ml. Although yields approaching these are obtainable by medium perfusion [6], they have not been achieved in batch cultures. This suggests that the limiting factor for improvements in cell yield in batch cultures is *not* the surface area for growth. Rather, some other yet unknown factors are responsible for growth limitation. For this reason a full exploration of the growth limiting factors should be given high priority so that microcarrier technology can reveal its full potential.

5.2 Seeding Density Effects

The inoculation of anchorage-dependent cells into a suspension of microcarriers is followed by several distinct stages:

a) cell to bead attachment
b) cell spreading on the bead surface
c) the growth of cells on each microcarrier to confluence

Once cell to bead attachment has occured, the transfer of cells from one microcarrier to another is not generally observed. Thus the cells that grow on each microcarrier can be considered as an independent colony.

Because of this independence of each microcarrier it is important to ensure an adequate initial distribution of inoculated viable cells throughout the population of microcarriers. This minimises the number of "empty" beads during culture and ensures maximum use of the surface area available for growth.

An analysis of cell attachment can be attempted by the use of the Poisson distribution equation [39]. The application of this equation assumes a probabilistic phenomenon of random attachment of cells to microcarriers following inoculation.

$$P = \frac{e^{-\lambda}\lambda^{n}}{n!} \quad (1)$$

where P = probability of a specific number (n) of cell hits per microcarrier.
where n = number of cell hits per microcarrier
where λ = ratio of cells per microcarrier in culture at time zero.

The validity of the use of the Poisson equation was assessed by Butler and Thilly [39] from experimental observations of the distribution of MDCK cells on DEAE-dextran microcarriers. Figure 4 shows a set of observations in which the proportion (P) of microcarriers carrying a specific number of cells (n) is plotted at various inoculation ratios of cells per microcarrier (λ). These observations were made by microscopic counting a few hours after inoculation. At this point cell attachment had occurred but not cell growth. For each value of λ the Poisson distribution (P vs n) was calculated mathematically and is also shown in Fig. 4.

The results of this work show that the distribution of the inoculated cells on microcarriers can be described by a skewed Poisson distribution. Comparison of empirical and theoretical plots in Fig. 4 shows a correlation which increases at higher λ values. The skewing of the distribution of the empirical values towards low cell/microcarrier attachment may be explained by the following:

a) There is a size distribution of any population of microcarriers. Cell attachment may favour the smaller microcarriers because of their slower movement in the stirred

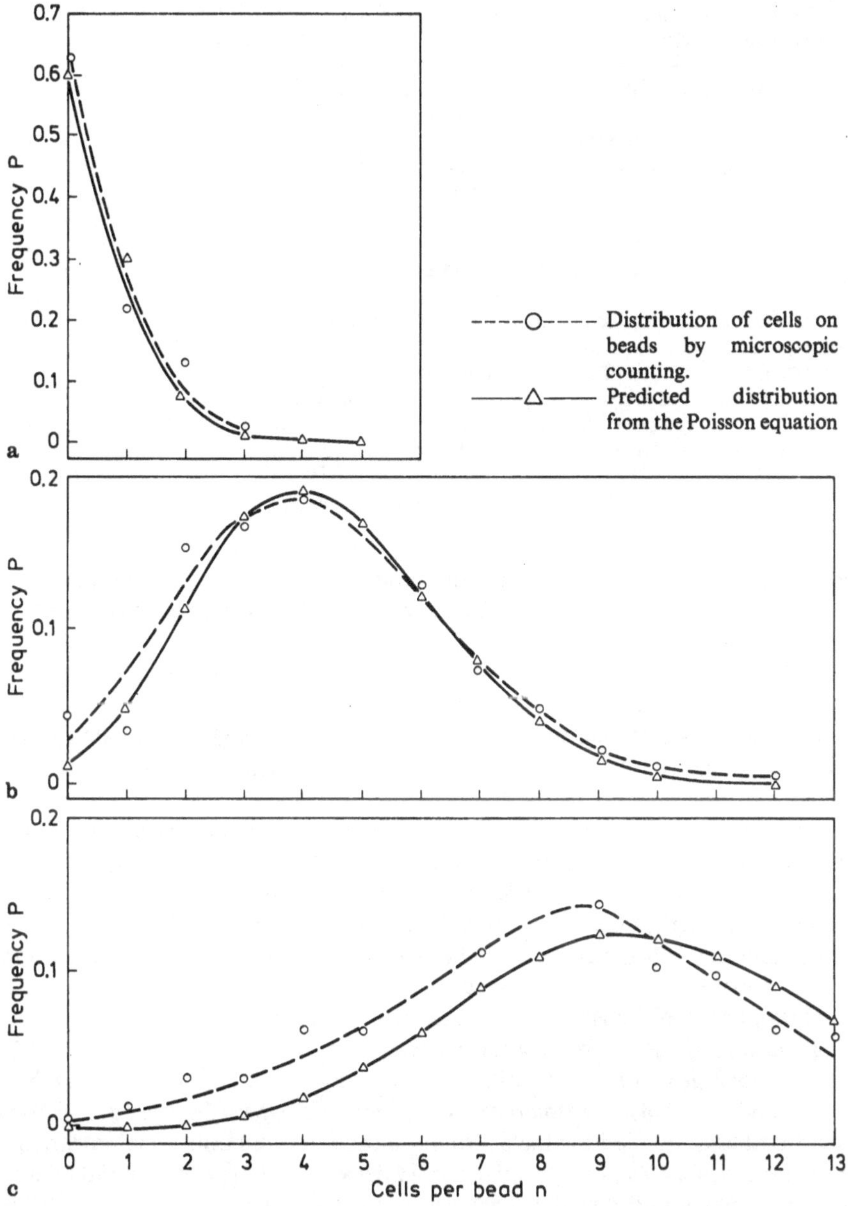

Fig. 4a—c. Distribution of cells on microcarriers at varying cell per bead inoculation ratios [40]. The MDCK cells were counted on microcarriers 3 h after inoculation by microscopic examination of samples treated with crystal violet (0.1 %) in PBS. The initial cell per bead inoculation ratio, λ was **a** 0.5, **b** 4.4 and **c** 9.9

Table 3. The effect of cell-to-bead inoculation ratios on number of unoccupied beads in culture [39]

Initial cell per bead ratio (λ)	Proportion of unoccupied beads counted[a]	Probability of zero hits per bead[b]
1.6	0.5	0.45
3.2	0.2	0.20
6.4	0.1	0.04
9.5	none detected	0.008

The results were taken from 100 ml microcarrier cultures of MDCK cells. The microcarrier concentration was maintained at 5 g L^{-1} whereas the inoculation concentration of cells was varied.

[a] These proportions were determined by microscopic examination of samples taken after cell growth to 1.3×10^6 cells per ml;

[b] These values were calculated from the Poisson equation

culture. This would cause an uneven cell distribution which may result in some larger beads not having attached cells. This latter phenomenon has been observed experimentally by the author.

b) The cell viability and cloning efficiency may not be 100% and there is a need to make an allowance for this in any calculations.

c) Some of the cells attached to microcarriers may be masked during microscopic counting. This may result in an underestimation of the number of cells attached to microcarriers.

The most useful result to be obtained from the analysis of cell distribution is the critical cell to microcarrier inoculation ratio necessary to obtain a negligible proportion of unoccupied ("empty") microcarriers in culture. From both empirical results and by the application of the Poisson equation the critical ratio for the MDCK cells used in the experiments described was shown to be >7 (Table 3). Thus a cell to microcarrier inoculation ratio above the critical value ensures:

a) a low proportion of unoccupied microcarriers in culture ($<5\%$).

b) maximum use of the available surface area by the cells.

A similar analysis of cell distribution was made by Hu et al. [8] for the growth of human diploid fibroblasts on microcarriers. They confirmed the use of a skewed Poisson distribution to describe the distribution of inoculated cells on microcarriers. Their mathematical analysis of the relationship between cell distribution and relative microcarrier size showed that the slight deviations from the normal Poisson distribution could be partly explained by preferential binding to small beads. They also showed that at a cell to bead inoculation ratio below the critical value — which was determined as 6 in this case — the final cell yield was reduced. Results showed that at a bead concentration of 5 mg ml^{-1} a cell inoculum concentration above 3×10^5 cells per ml was required to obtain the maximum cell yield of $1–2 \times 10^6$ cells per ml. At cell to bead inoculation ratios below the critical value and at a fixed microcarrier concentration the final cell yield can be shown to be directly related to the cell inoculation concentration. This is shown for both MDCK cells [39] and human diploid fibroblasts [8]

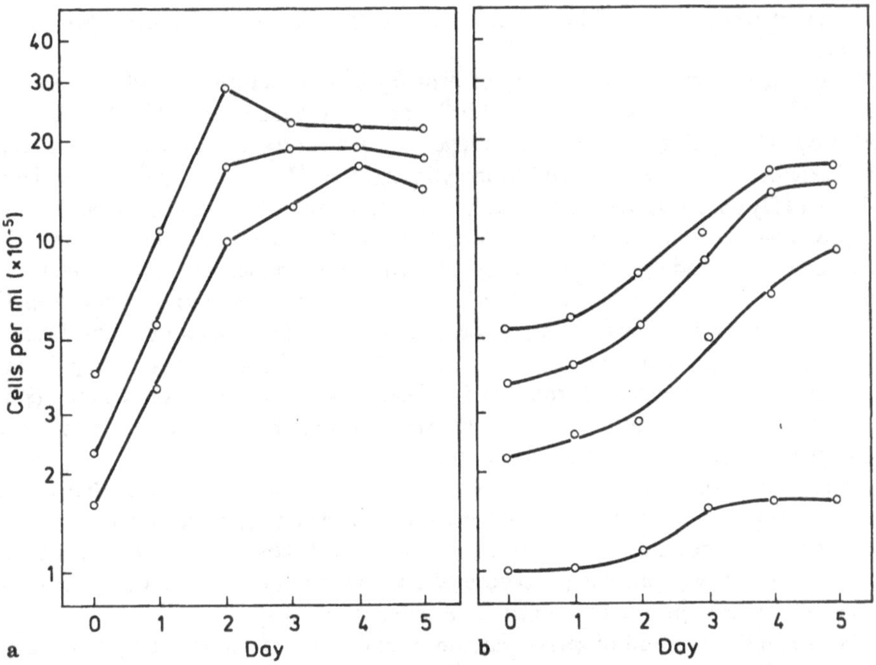

Fig. 5a and b. Growth of cells at varying cell per bead inoculation ratios. **a**, Canine epithelial cells (MDCK) grown in 100 ml cultures on DEAE-dextran microcarriers at 5 g L^{-1} [39], **b**, Human diploid fibroblast (FS-4) cells grown on DEAE-dextran microcarriers at 5 g L^{-1} [8]

in Fig. 5 Microscopic examination of the culures showed that the lower cell yields could be ascribed to higher proportions of "empty" microcarriers.

5.3 Scale-up

The stirred tank reactors (STR) traditionally used for microbial suspension cultures require only minimal modification for use for microcarrier cultures. The modifications include the provision of a larger impeller to maintain the microcarriers in suspension at low stirring speeds and also the avoidance of any internal edges in the fermenter which may create non-stirred spaces. The latter problem can be solved by the use of round-bottomed vessels. Such stirred tanks for microcarrier cultures have been built at least up to 1,000 litres [41, 42] with other systems up to 10,000 litres (unpublished).

Culture scale-up is generally performed in ten-fold increases. Thus the cells from one culture may be used to inoculate a culture with a ten-fold increase in volume. This assumes a cell multiplicity (= final cell yield per initial cell inoculum) of 10 which can be routinely obtained in microcarrier cultures.

With this type of scale-up operation it would be convenient to seed a vessel containing fresh medium and new microcarriers with confluent microcarriers isolated from the previous growth. However, the difficulty of this approach is that, as yet, it has not been found routinely possible to allow cells to transfer from one microcarrier to another.

Ryan et al. [43] reported the ability of primary bovine endothelial cells to transfer from confluent to fresh beads in roller bottles and also that cells could be detached by vortexing. They pointed out the advantages of such a procedure for cell harvesting that does not necessitate the use of proteolytic enzymes. However, it may be that these cells are easily detached from the beads because of the reported presence of peptidases and proteolytic enzymes on the cell surface. Other work has reported that only mitotic cells may be removed from microcarriers at high spinner velocities [44]. Burke et al. [35] reported that HeLa cells can remain attached to various microcarrier types at high shear forces. On a small scale, bead to bead transfer has been shown for fibroblastic cells statically held in culture dishes [29]. Crespi and Thilly [45] showed that two epithelial cell lines could be transferred from bead to bead in spinner flasks in a medium containing a low concentration of calcium: However, this method of cell transfer has not found general application for other cell lines.

Despite the papers cited above, a widely applicable method which allows cell transfer from bead to bead has not yet been devised. Such procedures might only be possible for cell lines that are not strongly attached to beads [42]. However, the development of a widely applicable procedure of this type would be a major step forward for industrial scale processes using microcarrier technology.

The alternative method of serial scale-up of microcarrier cultures is by enzymatic removal and cell harvesting from each culture. The treatment necessary varies with the nature of the microcarrier substratum. Trypsin and EDTA have been used for many years for the removal of cells from glass and plastic based vessels such as Roux bottles and T-flasks. The combined effects of proteolysis and metal chelation serve to remove the factors involved in cell-substratum attachment. A similar treatment can be used for cell detachment from glass-coated beads [14] which can then be re-used after washing. Re-attachment of such cells to new or cleaned microcarriers appears to be efficient for a number of cell lines tested [14,46] despite reported adverse effects of proteolytic enzymes on the membrane surface of cells [47,48].

The cell attachment to gelatin surfaces occurs by specific receptors and the linkage can be broken by the enzyme, collagenase. In the case of microcarriers composed completely of gelatin this will cause the beads to dissolve [11]. A similar complete removal of beads can be accomplished by centrifugation of the fluorocarbon fluid beads [30].

The removal of cells from DEAE-dextran microcarriers has proved rather more difficult. Scanning electron microscopic analysis has shown that this may be due to a greater cell to substratum contact compared to the long, slender filipodia observed on glass attachment [14]. However, Hu et al. [38] have shown efficient detachment and serial propagation of DEAE-dextran microcarrier cultures using a trypsin/EDTA mixture at high pH. A pH of 8.7 was shown to reduce the surface charge of the microcarrier and so ease cell detachment. Cell/bead separation was affected by column filtration. This detachment procedure was shown not to affect viability or product formation of human fibroblast or Vero cells and may be applicable to industrial scale-up. On the industrial scale, the process of cell trypsinisation may be aided by an apparatus that allows sterile transfers and mechanical agitation for cell detachment from microcarriers [49].

6 Nutrient and Oxygen Limitations

At the present state of technology, the microcarrier concentration of batch cultures can be increased sufficiently high so that surface area does not limit the growth of the cells. Thus, at the stationary phase in these cultures which typically occurs at 2×10^6 cells per ml, there still is surface area available for further cell growth. In this situation, the growth limitation must be due to the deterioration of the cellular micro-environment, necessary for continued cell growth. This limitation may be narrowed to either nutrient (including oxygen) supply or inhibitor accumulation. The precise factor that causes cells to stop growing in culture may vary with the cell line, media and growth conditions. However, an identification of these factors may be a step forward to enabling changes which will allow growth to much higher densities than those normally obtained.

6.1 Energy Source

The carbon energy sources provided in the culture media and their relative concentrations can influence the flux through the intracellular metabolic pathways [50]. This can affect the energy state of the cell and the relative amounts of waste metabolites produced (notably, lactic acid and ammonia).

6.1.1 Glucose

The carbohydrate energy source typically used for cell culture is glucose [51] at 5 to 20 mM. Although the glycolytic breakdown of glucose for energy metabolism is well recognised, this has not been found to be essential in cell culture if a sufficient supply of glutamine and pyrimidine nucleosides is provided [52]. At low concentration (< 25 μM) the primary metabolic function of glucose is to provide ribose for nucleic acid synthesis [53,54]. Media glucose levels < 20 μM causes cell detachment from microcarriers [53]. At higher glucose concentrations (20 mM) much of the carbon is converted to lactic acid which is secreted by the cell [54,55,56,57]. Only a small fraction of the glucose taken up by the cells may be metabolised by aerobic glycolysis. The substantial conversion of glucose to lactic acid is of concern because of the inefficient use of the carbon energy source compared to complete aerobic breakdown via the TCA cycle. This phenomenon is probably due to the Crabtree effect which is recognised as the inhibition of aerobic metabolism by high concentrations of sugars [58,59,60,61]. The effect has been explained by the competition for ADP between pyruvate kinase and mitochondrial oxidative phosphorylation [62]. The most efficient use of glucose as an energy source may be achieved by continuous feeding of a batch culture but maintaining the glucose at a constant, low concentration [53].

The substitution of glucose by fructose or maltose has been shown to reduce lactic acid production and pH changes in the culture of MDCK cells grown on microcarriers [55] and in non-microcarrier cultures by substitution by galactose [63]. This effect is probably related to a lower specific uptake of the substituted carbohydrates which may be preferentially metabolised via the pentose phosphate pathway [52].

The specific rate of glucose uptake can vary with the pH of the culture [64], stage of growth [59,62] and with the cell line by a factor of 10 [65]. The glucose content of the

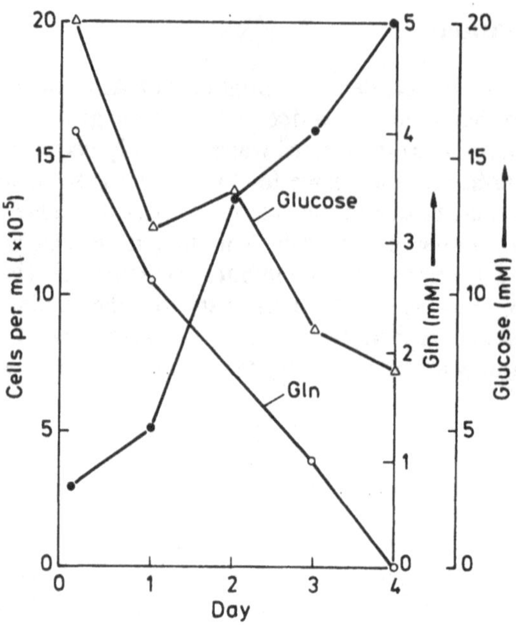

Fig. 6. Consumption of glucose (△) and glutamine (○) during cell growth (●) in microcarrier cultures. BHK cells were grown in 100 ml cultures with DEAE-dextran microcarriers at 2.5 g L^{-1} Glucose and glutamine were measured enzymatically [66]

medium was not completely depleted by the growth of BHK cells on microcarriers even at a reduced initial concentration (10 mM) [66] (Fig. 6). This suggests that in this system the glucose supply is not a limiting growth factor for cell yields.

6.1.2 Glutamine

Glutamine is normally included in culture medium at much higher concentrations than any other amino acids (0.7—5 mM). The role of glutamine has been recognised as a major energy source in cell culture media [57] although its other metabolic functions include lipid biosynthesis [67], use as a precursor for the biosynthesis of the amino acids, aspartic acid, proline and asparagine [68] and as a regulatory component of DNA replication [69]. It is also used with the other amino acids present in the medium for protein biosynthesis.

The metabolic pathways followed by glutamine in cultured human diploid fibroblasts were studied by Zielke et al. [53, 54, 70]. These cells showed a reciprocal regulation in the utilisation of glucose and glutamine [53]. Thus, increased glutamine utilisation was observed at low medium concentrations of glucose which was required to be at 25 μM for maintenance of normal growth rates. Glutamine oxidation contributed 30% of the cellular energy requirements in standard media [53]. This metabolic oxidation involves a two stage deamination of glutamine to glutamic acid and α-ketoglutaric acid which can be completely metabolized to carbon dioxide by the TCA cycle [71, 72, 73]. The conversion of some glutamine to lactic acid has also been demonstrated [70]. Radioactive ^{14}C-labelling experiments showed the conversion of 4-carbon TCA cycle intermediates into 3-carbon glycolytic intermediates and implicates the importance of the enzyme, phosphoenol pyruvate carboxykinase [70].

The specific rate of utilisation of glutamine is rapid compared to other media components and complete depletion has been shown in batch cultures of MDCK [39] and BHK [66] cells grown to confluence on microcarriers (Fig. 6). Complete media depletion of glutamine was also shown for suspension cultures of mouse L-cells [74]. However the specific cellular consumption of both glutamine and glucose is dependent upon their initial concentrations in the medium but is not necessarily related to the growth rate or the final cell yield [53,66].

The optimum initial concentration of glutamine for BHK cells on microcarriers was established as 4 mM [66]. Further addition of glutamine leads to excessive accumulation of ammonia which may be toxic [66]. This latter problem may be overcome by substitution of glutamine by other chemically similar carbon sources such as glutamic acid. However a process of cellular adaptation must be followed before this is possible [75] and the full implications of this for microcarrier cultures have not yet been established.

6.2 Amino Acids

Amino acids are normally added as defined components in cell growth media. However, the addition of undefined supplements such as serum, tryptose phosphate broth (TPB) or lactalbumin hydrolysate (LH) can raise further the amino acid content of the medium [76].

The consumption of amino acids from medium supporting the growth of MDCK cells on microcarriers was studied by Butler and Thilly [39] (Table 4). The behaviour of the amino acids during cell growth was divided into 3 types. The first group included 8 amino acids which were rapidly and completely consumed during growth. The second group included 5 amino acids which showed limited depletion. The third group — 2 amino acids were released into the medium by the cells. The rapid consumption of the branched chain amino acids — leu, ile & val — is a phenomenon observed in MDCK [39] and other cell lines including human diploid fibroblasts [78], mouse myeloma cells [79] and BHK cells [65]. The sulphur containing amino acids — met & cys — are also found amongst the rapidly consumned [39,78]. The release of specific amino acids into the culture medium during cell growth has been commonly found particularly for glycine and alanine [39,65,80]. This is likely to be precipitated by a build up of excess intracellular ammonia which is sequestrated by binding to available carbon metabolites.

The primary importance of the amino acids present in culture media is to act as precursors for protein biosynthesis. However, studies on cellular protein production in BHK cells in suspension culture showed that 30% of the amino acid uptake was not incorporated into cellular protein [76]. This suggests that those amino acids may be deaminated and incorporated into energy metabolism.

The complete depletion of many essential amino acids during cell growth in batch cultures must contribute to a limitation of the final cell yield in microcarrier cultures. One solution to this problem may be to selectively feed the amino acids at rates that reflect their cellular utilisation. An alternative procedure is to allow complete medium perfusion at a rate corresponding to the cellular demand. Table 4 shows the changes of the amino acid content of a culture of MDCK cells grown on microcarriers in a

Table. 4. The consumption of amino acids by MDCK cells grown on microcarriers [77]

Amino acid	Final concentrations as a % of original	
	Unperfused culture	Perfused culture
Arg	0	53
Cys	11	50
Gln	0	25
Ile (a)	0	77
Leu	0	77
Met	0	28
Val	0	76
His	47	60
Lys	66	77
Phe (b)	33	75
Thr	54	—
Tyr	54	70
Ala	100	3000
Glu (c)	—	250
Gly	133	105

Complete amino acid analysis was performed on the culture media at the end of cell growth in perfused and unperfused cultures of MDCK cells. The values presented are amino acid concentrations expressed as perentages of the initial values determined in the culture medium. The final cell concentration in the perfused cultures was 9×10^6 cells per ml and the perfusion rate was 3 volumes per day [6]. The final cell concentration in unperfused cultures was 3×10^6 cells per ml [39].

Patterns of amino acid utilisation in unperfused cultures fall into 3 catagories — a, b and c

perfusion system [6]. In this system, the concentrations of those essential amino acids which were completely depleted in batch cultures, were maintained at 50% of original levels by perfusion at 3 volumes per day. Clearly a finer control of the cellular needs could be met by increasing the perfusion rate in response to the increase in cell concentration.

6.3 Serum and Hormone Supplements

The growth of cells in microcarrier cultures normally requires the presence of 5 to 10% serum in the medium. The purpose of the serum includes the following:

a) To promote attachment of cells to the microcarrier surface by providing specific glycoproteins.

b) To promote cell growth by providing growth factors only some of which are well defined.

c) To protect the cells from mechanical damage.

Despite the need for such serum for optimal growth of cells there are a number of disadvantages involved in the use of serum:

a) Serum is chemically undefined and its hormonal and growth promoting components vary considerably in content between types and even between batches of the same type [78,81]. This can cause problems of reproducibility of culturing.

b) The addition of serum causes a considerable increase in the overall protein content of the culture medium. This may increase the difficulty and cost of purification of the extracellularly released products i.e.: the downstream processing of the desired biologicals.

c) Serum is the single most expensive component of culture medium and this is particularly significant in dealing with industrial processes > 10^3 litres.

d) Serum is vulnerable to infection from viruses or mycoplasmas derived from the donor animal.

For the above reasons, a number of attempts have been made to formulate reduced serum or serum-free media suitable for the growth of cells in vitro [82,83,84]. For some cell lines, a period of gradual adaptation using successive decreases in serum concentration is required [85]. The role of serum in microcarrier cultures can be differentiated into requirements for cell attachment and cell proliferation. A high serum concentration is often required at the early stage of culture to provide those factors that allow efficient cell to bead attachment. However, the serum content need not be constant throughout the cell culture cycle. Replenishement or perfused media may be of lower serum content but may still allow efficient growth rates. The serum content can be reduced even further at the stationary phase at which time it has been reported that a low serum content (< 0.5%) can reduce the shedding of a confluent monolayer from microcarriers [86].

Most serum-free formulations require supplementation with hormones and growth factors. The hormone supplements that can typically replace serum include fibronectin, transferrin, insulin, steroids and polypeptide growth factors. Some of these formulations can be used or adapted for use in microcarrier cultures [87]. For the growth of MDCK cells on microcarriers, an increase in the concentration of some of the hormones and the addition of cAMP to a previously used serum-free formulation was found desirable [9]. In some cases, the addition of high molecular weight components such as serum albumin or Ficoll to serum-free formulations may be necessary to protect the cells [88]. Subculture by trypsinisation can be a problem in serum-free medium due to the absence of inhibitors but this may be overcome by the use of soybean trypsin inhibitor.

6.4 Oxygen

The supply of oxygen to a cell culture is a critical parameter for attaining high densities and for scaling-up to large volumes. In small scale cultures the cellular oxygen demand is normally satisfied by diffusion of oxygen from the head space of the culture vessel. The oxygen transfer rate by this process can be expressed by the following equation:

$$OTR = K_L A(C^* - C_L) \tag{2}$$

where OTR = oxygen transfer rate (mMol O_2 h^{-1})
where K_L = rate coefficient
where A = surface area exposed to gas phase
where C* = saturated concentration of O_2 in the medium
where C_L = actual concentration of O_2 in the medium

In order that oxygen does not become a limitation to cell growth, the oxygen transfer rate (OTR) must satisfy the oxygen utilisation rate (OUR) of the cells in culture. The oxygen utilisation rate can vary according to the cell line from 0.05 mMol l^{-1} h^{-1} [89, 90] to 0.6 mMol l^{-1} h^{-1} [91] for a culture at 10^6 cells per ml. Katinger and Scheirer [92] assumed on OUR of 0.063 mMol l^{-1} h^{-1} and a determined transfer coefficient of 0.63 μMol O_2 cm^{-2} h^{-1} to calculate the culture volume at which the rate of oxygen supply would become limiting. Table 5 shows that, assuming a cell density of 10^6 cells per ml, the OUR exceeds the OTR at a culture volume of 10 litres. Therefore, at culture volumes above 10 litres (for 10^6 cells per ml) or cell densities above 10^6 per ml (for 1 litre cultures), an alternative to oxygen diffusion from the head space of air is required. This conclusion is confirmed by Fleischaker and Sinskey [90] who found that 5 L is the approximate limit for the growth of human diploid fibroblasts in microcarrier cultures without additional aeration.

Several solutions have been suggested to satisfy this greater oxygen demand in cultures at high cell densities and culture volumes:

a) The head space could be filled with oxygen instead of air. This can increase the critical volume or the critical cell density by a factor of 3.5 [1]. Clark and Hirtenstein [93] used a system for microcarrier cultures in which medium was pumped to a second vessel for oxygen equilibration from the head space.

b) Air sparging which is commonly performed in microbial cultures is unsuitable since it causes damage to animal cells and foaming of the medium [94]. This is a particular problèm for microcarrier cultures, as the microcarriers themselves tend to concentrate in the foam [90]. However, a caged aeration system which produces sufficiently small gas bubbles, can reduce these problems of cell damage and medium foaming [95]. Such a system was found suitable for the aeration of microcarrier cultures up to 10 litres.

Table 5. Oxygen demand versus surface aeration at different reactor volumes [92]

Culture volume (litre)	Head space area (cm^2)	Oxygen supply from head space (mMol h^{-1})	Cellular O_2 demand (mMol h^{-1})
1	100	0.063	0.063
10	500	0.313	0.625
100	2500	1.56	7.81

These calculations are based on the follwing assumptions: A culture with liquid height (H)/diameter (D) = 1; Cell concentration in culture = 10^6 cells per ml and the following experimentally determined values: Head space areation = 0.63 μMol O_2 cm^{-2} h^{-1}; Oxygen utilization rate = 0.063 mMol O_2 L^{-1} h^{-1}

c) Oxygen can be supplied by diffusion through thin-walled silicone tubing introduced into the liquid medium. This method has been used to supply oxygen to a 10 litre microcarrier culture of human diploid fibroblasts [90, 96]. However, this system requires 0.5 m of tubing per L of culture which could become unwieldy for scale-up [1].

As well as considering the method of oxygen supply, it may also be important to maintain the optimum dissolved oxygen tension (DOT) for cell growth. From the data available, considerable variability is shown in the sensitivity of cell lines to DOT. In some cases little effect is found upon the growth rate or maximum cell density over a wide range of DOT values [97] although in other examples sharp optima are found often at relatively low DOT values [98, 99, 100, 101].

Of particular significance to microcarrier cultures is the finding that cell-surface attachment is far less dependent on DOT than cell growth. However, low oxygen tensions are generally favoured at low cell densities, whereas optimal DOT for exponential growth can be higher [94, 99, 100, 102, 103]. This suggests the desirability for the control of oxygen tensions in microcarrier cultures of sensitive cells from low to higher tensions as cell growth progresses [5]. In many systems now available this optimal level may be maintained by use of a sterilisable oxygen probe continuously monitoring the DOT and with a computerised feedback system controlling the oxygen supply.

7 Inhibitor Limitations

The micro-environment of cells during growth in batch cultures can deteriorate by the extracellular release of compounds or by the chemical degradation of media components. The two compounds which are well-known and well-defined in this respect are — lactic acid and ammonia — although other hitherto unknown compounds may also be important in media deterioration. Such changes in the composition of the medium may lead to growth limitation which may only be eliminated by the removal of the accumulated compound.

7.1 Lactic Acid

Lactic acid is released into culture medium as the product of anaerobic glycolysis resulting from carbohydrate metabolism [54, 55, 56, 57] or to a more limited extent by glutamine metabolism [70]. The accumulation of lactic acid in the medium is recognised by the steady change in colour of the pH indicator during the cell growth cycle. This lowering in pH can lead to many adverse effects including reduced cell growth, reduced cell viability and cell detachment from microcarriers [5].

The accumulation of lactic acid can be minimized in a number of ways:

a) The supply of glucose to the cells can be controlled by supplementation of media with a low level concentration (25—80 μM) followed by re-feeding [53] or by continuous batch feeding [1]. Low concentrations of glucose reduce the Crabtree effect which is an inhibition of aerobic metabolism.

b) The use of carbohydrate sources other than glucose e.g.: fructose [55] or galactose [63].

c) The use of biotin has been found to reduce lactate production in transformed cells [104].

The toxicity of lactate as opposed to its effect on pH was studied in the growth of hybridomas by Reuveny et al. [105]. They showed that 26 mM lactate was accumulated after complete cell growth in spinner flasks but that pH adjusted media containing up to 28 mM lactate did not cause growth inhibition. If this relatively low toxicity of lactate is applicable to other cell lines it may only be necessary to ensure adequate pH control of the culture to limit the adverse effects of lactate accumulation.

7.2 Ammonia

The accumulation of ammonia in culture medium results from a combination of chemical decomposition and metabolic deamination of glutamine. In buffered culture medium (pH 7.4) most of the ammonia is in the form — NH_4^+ — and at the levels normally found, it has little effect on the pH of the culture. The spontaneous decomposition of glutamine [106] at 37 °C can result in a rate of ammonia production of 0.1 mM per day [66]. This is however insignificant compared to the rate of metabolic deamination which results in the accumulation of 2—3 mM NH_4^+ in a complete growth cycle on microcarriers in 3 to 4 d [6,66,77]. The toxicity and inhibitory effects of this accumulated ammonia has been assessed in a number of experimental systems: —
a) MDCK cells grown in a perfusion system on microcarriers showed an increased cell yield of 10^7 cells per ml compared to 2×10^6 cells per ml obtainable by equivalent batch cultures. In both systems the ammonia concentration rose to 2.5 mM at the point of

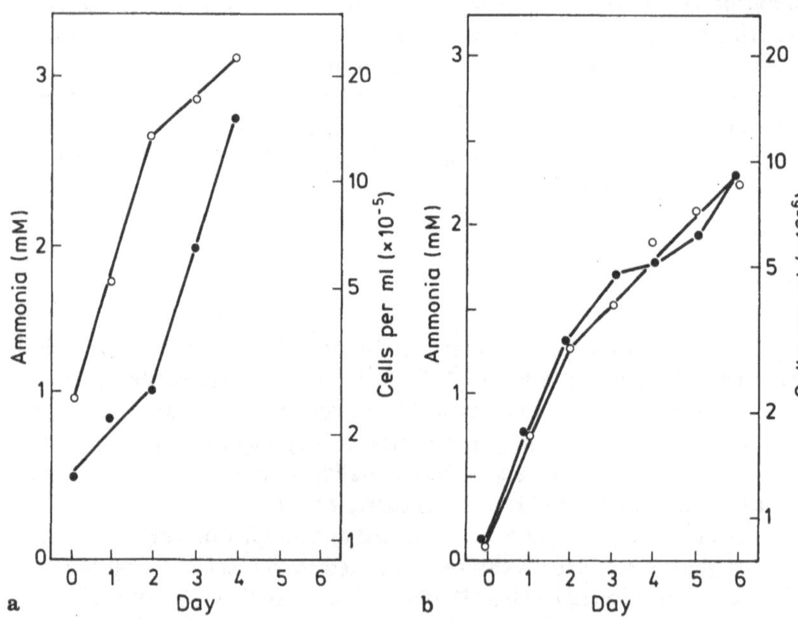

Fig. 7a and b. Accumulation of ammonia in the medium of microcarrier cultures during cell growth. Cell concentration (○) and ammonia concentration (●) were determined in: **a**, BHK cells [66] grown in batch cultures (100 ml) on DEAE-dextran microcarriers (2.5 g L^{-1}); **b**, MDCK cells [6] grown in perfusion cultures (1 L) on DEAE-dextran microcarriers (7.5 g L^{-1}) with a rate of perfusion of 3 volumes per day

cessation of cell growth [6]. Similar levels of accumulated ammonia are found in BHK cultures[66] (Fig. 7).
b) The dilution of culture medium (1 : 1v/v) with phosphate buffered saline (PBS) did not significantly alter the cell yield obtainable by medium perfusion. This suggests that the rate of removal of inhibitors from the culture is a more important factor in obtaining a high cell yield in this system than the supply of nutrients [6].
c) The supplementation of culture medium with ammonium chloride at a concentration > 1 mM caused significant growth inhibition of BHK cells [66] (Table 6). The inhibition was greater in microcarrier cultures. The concentration of NH_4^+ that normally accumulates in batch cultures at the cessation of cell growth is 2–3 mM.

Table 6. Inhibition of cell growth by ammonium chloride [66]

Added NH_4^+ (mM)	Productivity (10^6 cells per ml)	
	Microcarrier cultures^a	Cultures in small flat bottles^b
0	11.81	8.64
1	4.86	7.39
2	0	6.28
3	not determined	1.20

The growth of BHK cells was determined over 3 d in cultures containing concentration of added ammonium chloride
^{a} Bells were grown on DEAE-dextran microcarriers (2.5 g L^{-1}) in 40 ml cultures
^{b} Cells were grown in small flat bottles (56 g) in 15 ml culture

d) A dilution of 1 : 1v/v of culture medium by PBS caused only a 15% reduction in the final cell yield of BHK cells [66]. This suggests that nutrient supply to the cells is not the major factor controlling cell yield.
e) Hybridomas grown in batch cultures were observed to produce ammonia concentrations of 4.5—5.5 mM. The toxic effects of ammonium supplementation to the growth media were observed > 2 mM [105].
f) A decrease in ammonia accumulation and an increased cell yield was shown for hybridomas by use in culture of a dialysis tube containing an ammonium ion absorbent [107].
g) Ammonium ion levels as low as 70 μM have been shown to inhibit the effect of a range of viruses on cells grown in culture [108].

The experimental evidence cited above points to the importance of ammonia accumulation in the limitation of cell yields in culture. A number of approaches are possible to attempt the limitation of such accumulation: —
h) The perfusion of media through cultures at rates that prevent ammonium accumulation to toxic levels. Butler et al. [6] perfused MDCK cells on microcarriers at 3 volumes per day and found a 2.5 mM ammonium ion accumulation in 6 d at a final cell yield of 10^7 cells per ml. Faster rates or differentially increasing rates corresponding to cell growth could improve these yields. The problem of this approach, particularly for an industrial process, is the increased media cost.

i) The use of alternative carbon substrates that limit ammonia production. The reciprocal regulation of glucose and glutamine suggests that the rate of glutamine utilisation can be decreased by an increased supply of glucose [53]. However, this may result in undesirable levels of lactic acid. Cells may be adapted to utilise glutamic acid [75]. The implications of this approach to microcarrier cultures and its general applicability require further exploration.
k) The continuous specific removal of ammonium ions from culture. Dialysis was attempted by Iio et al. [107] and led to encouraging results.

However, the problem of specificity must be considered. Such a procedure may cause the removal of a range of small molecular weight compounds from the medium.
l) By specific additions or continuous batch-feeding of glutamine which is maintained at low concentrations (< 0.2 mM) in the medium [1,109]. This approach has been found to reduce the ammonia production from assimilated glutamine.

8 Products from Cell Growth on Microcarriers

The increasing range of commercially useful biologicals that can be produced from animal cells [110] has been a major impetus for developments in cell culture technology. Many of these biologicals are produced from anchorage-dependent cells and microcarrier culture systems offer an excellent means of scale-up [5,111]. In particular, the advantages offered by microcarriers for such production include:
a) the ease of control and operation of a homogeneous unit process.
b) the ease of separation by sedimentation of microcarriers from culture so that the normally extracellularly released product can be isolated from the supernatant.

As the full range and extent of the production of biologicals from microcarrier grown cells has been well reviewed elsewhere [5,24,111,112,113], only examples of the major industrial products will be mentioned here. These include:

8.1 Vaccines

The production of viral vaccines from animal cell substrates was developed on a large-scale in the 1950s as a multiple culture process. The realisation of the possibilities for adaptation to a unit microcarrier culture came readily with the development of the technology and and has resulted in culture systems for the production of Foot-and-Mouth Disease [114] and Poliomyelitis [41,49] vaccines of several hundred to a thousand litres. Reports have shown that virus yields per cell from such microcarrier cultures can be greater than using multiple culture bottles [112,115].

Vaccine production involves a two stage process of cell propagation followed by virus multiplication. The environmental conditions for each stage are often different and critical in order to maximize yields [42,116]. However, regimes have been developed for rapid changes in conditions and media where necessary so that the down-time of the production process is minimized [49,113].

8.2 Interferon

The production of β-interferon for potential clinical use can be induced from human diploid fibroblasts which can be readily grown in large-scale microcarrier cultures [117,118,119]. Following cell growth, a process of induction ("Superinduction") is required to allow interferon production from cells [120]. This induction process, like the viral propagation stage for vaccine production, requires changes of media and environmental conditions that are different from cell growth [121, 122]. However, such changes can be made relatively easily in controlled microcarrier cultures.

The two examples given above highlight products which are already being produced on an industrial scale from microcarrier cultures up to 10^3 L. The full list is extensive and is likely to increase with the realisation of the value of extracellularly released products from eukaryotes [123] and the possibilities of useful genetically engineered animal cells.

9 Conclusion

Microcarrier culture systems have found a wide application for the production of useful biologicals, examples of which have been highlighted here. In order to increase the product yield from such processes, either the specific productivity per cell may be improved or the total cell yield per unit culture may be improved. The prospect of the latter has potential benefits for all processes involving animal cell cultures.

Of the range of microcarrier types available, certain cell lines may be preferentially grown on particular microcarriers. However, most if not all animal cells capable of growing as monolayers on culture dishes are capable of growth on microcarriers. The kinetics of such growth has been well studied and certain parameters such as microcarrier concentration and initial cell to bead ratios need to be considered in order to maximize cell yields. Having considered these parameters, it is then possible to grow cells in a system where surface area is not the limiting factor in determining the final cell yield.

In such a system, improvements in cell yield may be made by medium perfusion but the increased utilization of medium can cause prohibitive costs. To allow improvements in cell yield in batch cultures a careful study is required of the media composition and micro-environment of the cells. Although each cell line studied may have different sensitivities to such factors as nutrient limitation or inhibitor concentration each of these problems is likely to be met at some point before the maximum theoretical yield can be attained at high microcarrier concentrations.

The solution to these problems lies in the fine control of the media composition and culture environment during cell growth. In some cases, such as oxygen supply, the problem is well defined and can be controlled. In most cases, the interaction of the cell with its micro-environment is poorly understood and the complexity of the changes in culture media does not allow easy control. However, it is anticipated that in the next few years a fuller understanding of control will lead to considerable improvements in cell yield which will benefit the economics of industrial scale processes using microcarrier cultures.

10 Nomenclature

ADP	adenosine diphosphate
ATP	adenosine triphosphate
cAMP	cyclic adenosine monophosphate

DEAE	diethyl amino ethane
DMAP	dimethyl amino propane
DOT	dissolved oxygen tension
EDTA	ethylene diamine tetra-acetic acid
LH	lactalbumin hydrolysate
MDCK	Madin Darby canine kidney
OTR	oxygen transfer rate
OUR	oxygen utilisation rate
PBS	phosphate buffered saline
TCA	tricarboxylic acid
TPB	tryptose phosphate broth

11 References

1. Glacken, M. W., Fleischaker, R. J., Sinskey, A. J.: Trend in Biotech. *1*, 102 (1983)
2. van Wezel, A. L.: Nature *216*, 64 (1967)
3. Levine, D. W., Wong, J. S., Wang, D. I. C., Thilly, W. G.: Somatic Cell Genetics *3*, 149 (1977)
4. Levine, D. W., Wang, D. I. C., Thilly, W. G.: Biotech. Bioeng. *21*, 821 (1979)
5. Pharmacia trade publication: Microcarrier Cell Culture: Principles, Methods. Uppsala, Sweden 1981
6. Butler, M., Imamura, T., Thomas, J., Thilly, W. G.: J. Cell Sci. *61*, 351 (1983)
7. Feder, J., Tolbert, W. R.: Int. Biotech. Lab. *3*, 40 (1985)
8. Hu, W. S., Meier, J., Wang, D. I. C.: Biotech. Bioeng. *27*, 585 (1985)
9. Crespi, C. L., Imamura, T., Leong, P.-M., Fleischaker, R. J., Brunengraber, H., Thilly, W. G., Giard, D. J.: ibid. *23*, 2673 (1981)
10. Gebb, C., Clark, J. M., Hirtenstein, M. D., Lindgren, G., Lindskog, U., Lundgren, B., Vretblad, P.: Develop. Biol. Stand. *50*, 93 (1982)
11. Paris, M. S., Eaton, D. L., Sempolinski, D. E., Sharma, B. P.: 34th Ann. Meet Tissue Culture Assoc. (1983)
12. Spier, R. E., Whiteside, J. P., Bolt, K.: Biotech. Bioeng. *19*, 1735 (1977)
13. Whiteside, J. P., Whiting, B. R., Spier, R. E.: Develop. Biol. Stand. *46*, 187 (1980)
14. Varani, J., Dame, M., Beals, T. F., Wass, J. A.: Biotech. Bioeng. *25*, 1359 (1983)
15. Reuveny, S., Silberstein, L., Shahar, A., Freeman, E., Mizrahi, A.: Develop. Biol. Stand. *50*, 115 (1982)
16. Reuveny, S., Silberstein, L., Shahar, A., Freeman, E., Mizrahi, A.: In Vitro *18*, 92 (1982)
17. Talbot, P., Keen, M. J.: Develop. Biol. Stand. *46*, 147 (1980)
18. Reuveny, S., Bino, T., Rosenberg, H., Mizrahi, A.: ibid. *46*, 137 (1980)
19. Johansson, A., Nielsen, V.: ibid. *46*, 125 (1980)
20. Maroudas, N. G.: J. Cell Physiol. *90*, 511 (1976)
21. Nielsen, V., Johansson, A.: Develop. Biol. Stand. *46*, 131 (1980)
22. Kuo, M. J., Lewis, C. Jr., Martin, R. A., Miller, R. E., Schoenfeld, R. A., Schuck, J. M., Wildi, B. S.: In Vitro *17*, 901 (1981)
23. Morandi, M., Bandinelli, L., Valeri, A.: Experientia *38*, 668 (1982)
24. Spier, R. E.: Adv. Biochem. Eng. *14*, 119 (1980)
25. Reuveny, S., Mizrahi, A., Kotler, M., Freeman, A.: Biotech. Bioeng. *25*, 469 (1983)
26. Hirtenstein, M., Clark, J., Lindgren, D., Vretblad, P.: Develop. Biol. Stand. *46*, 109 (1980)
27. Reuveny, S., Mizrahi, A., Kotler, M., Freeman, A.: Biotech. Bioeng. *25*, 2969 (1983)
28. Reuveny, S., Mizrahi, A., Kotler, M., Freeman, A.: Develop. Biol. Stand. *55*, 11 (1983)
29. Obrenovitch, A., Maintier, C., Sene, C., Boschetti, E., Monsigny, M.: Biol. Cell *46*, 249 (1983)
30. Keese, C. R., Giaever, I.: Science *219*, 1448 (1983)
31. Sherbet, G. V.: The biophysical characterisation of the cell surface. Academic Press 1978
32. Borysenko, J. Z., Woods, W.: Exp. Cell Res. *118*, 215 (1979)
33. Grinnell, F.: Int. Rev. Cytol. *53* 65 (1978)

34. Maroudas, N. G.: J. Theor. Biol. *49*, 417 (1975)
35. Burke, D., Brown, M. J., Jacobson, B. S.: Tissue Cell *15*, 181 (1983)
36. Fairman, K., Jacobson, B. S.: ibid. *15*, 167 (1983)
37. Tolbert, W. R., Feder, J. F.: Ann, Reports on Fermentation Processes *6*, 35 (1983)
38. Hu, W. S., Giard, D. J., Wang, D. I. C.: Biotech. Bioeng. *27*, 1466 (1985)
39. Butler, M., Thilly, W. G.: In Vitro *18*, 213 (1982)
40. Butler, M., Hassell, T., Rowley, A.: Proc. Symp. on Process Possibilities for Plant, Animal Cell Cultures: UMIST/Inst. Chem. Eng. (in press 1986)
41. Montagnon, B., Vincent-Falquet, J. C., Fanget, B.: Develop. Biol. Stand. *55*, 37 (1984)
42. van Wezel, A. L.: ibid. *55*, 3 (1984)
43. Ryan, U. S., Mortara, M., Whitaker, C.: Tissue, Cell *12*, 219 (1980)
44. Ng, J. J. Y., Crespi, C. L., Thilly, W. G.: Anal. Biochem. *109*, 231 (1980)
45. Crespi, C. L., Thilly, W. G.: Biotech. Bioeng. *23*, 983 (1981)
46. Varani, J., Dame, M., Rediske, J., Beals, T. F., Hillegas, W.: J. Biol. Stand. *13*, 67 (1985)
47. Cassiman, J., Brugmans, M., van den Berghe, H.: Cell Biol. Int. Reports *5*, 125 (1981)
48. Schor, S.: J. Cell Sci. *40*, 271 (1979)
49. van Wezel, A. L., van der Velden de Groot, C. A. M., van Herwaarden, J. A. M.: Develop. Biol. Stand. *46*, 151 (1980)
50. Morgan, M. J., Faik, P.: Bioscience Reports *1*, 669 (1981)
51. Eagle, H., Barban, S., Levy, M., Schulze, H. O.: J. Biol. Chem. *233*, 551 (1958)
52. Wice, B. M., Reitzer, L. J., Kennell, D.: ibid. *256*, 7812 (1981)
53. Zielke, H. R., Oz, P. T., Tildon, J. T., Sevdalian, D. A., Cornblath, M.: J. Cell Physiol. *95*, 41 (1978)
54. Zielke, H. R., Oz, P. T., Tildon, J. T., Sevdalian, D. A., Cornblath, M.: Proc. Nat. Acad. Sci. *73*, 4110 (1976)
55. Imamura, T., Crespi, C. L., Thilly, W. G., Brunengraber, H.: Anal. Biochem. *124*, 353 (1982)
56. Robinson, J. H., Butlin, P. M., Imrie, R. C.: Develop. Biol. Stand. *46*, 173 (1980)
57. Reitzer, L. J., Wice, B. M., Kennell, D.: J. Biol. Chem. *254*, 2669 (1979)
58. De Deken, R. M.: J. Gen, Microbiol. *44*, 149 (1966)
59. Zwartouw, H. T., Westwood, J. C. N.: Brit. J. Exp. Pathol. *39*, 529 (1959)
60. Chico, E., Olavarria, J. S., Nunez de Castro: Biochem. Biophys. Res. Comm. *83*, 1422 (1978)
61. Cristofalo, V. J., Kritschersky, D.: Proc. Soc. Exp. Biol. Med. *118*, 1109 (1965)
62. Gosalvez, M., Perez-Garcia, J., Weinhouse, S.: Eur. J. Biochem. *46*, 133 (1974)
63. Leibovitz, A.: Amer. J. Hyg. *78*, 173 (1963)
64. Birch, J. R., Edwards, D. J.: Develop. Biol. Stand. *46*, 59 (1980)
65. Arathoon, W. R., Telling, R. C.: ibid. *50*, 145 (1982)
66. Butler, M., Spier, R. E.: J. Biotech. *1*, 187 (1984)
67. Reed, W. D., Zielke, H. R., Baab, P. J., Oz, P. T.: Lipids, *16*, 677 (1981)
68. Levintow, L.: Science *126*, 611 (1957)
69. Zetterberg, A., Engstrom, W.: J. Cell Physiol. *108*, 365 (1981)
70. Zielke, H. R., Sumbilla, C. M., Sevdalian, D. A., Hawkins, R. L., Oz, ,P. T.: ibid. *104*, 433 (1980)
71. Lavietes, B. B., Regan, D. H., Demopoulas, H. B.: Proc. Nat. Acad. Sci. *71*, 3993 (1974)
72. Stoner, G. D., Merchant, D. J.: In Vitro *5*, 330 (1972)
73. Kovacevic, Z., Morris, H. P.: Cancer Res. *32*, 326 (1972)
74. Griffiths, J. B., Pirt, S. J.: Proc, R. Soc. (London) Ser B *168*, 421 (1967)
75. Griffiths, J. B.: J. Cell Sci. *12*, 617 (1973)
76. Mizrahi, A., Avihoo, A.: J. Biol. Stand. *4*, 51 (1976)
77. Butler, M.: Develop. Biol. Stand, *60*, 269 (1985)
78. Lambert, K., Pirt, S. J.: J. Cell Sci. *17*, 397 (1975)
79. Roberts, R. S., Hsu, H. W., Lin, K. D., Yang, T. J.: ibid. *21*, 609 (1976)
80. Polastri, G. D., Friesen, H. J., Mauler, R.: Develop. Biol. Stand. *55*, 53 (1984)
81. Esber, H. J., Payne, I. J., Bogden, A. E.: J. Nat. Cancer Inst. *50*, 559 (1973)
82. Barnes, D., Sato, G.: Cell *22*, 649 (1980)
83. Mather, J. P. (ed.) Mammalian cell culture — the use of serum-free hormone-supplemented media. Plenum Press 1984
84. Butler, M.: Serum-free media, in: Mammalian Cell Technology (Thilly, W. G., ed.) p. 91, Butterworths 1986

85. Giard, D. J., Fleischaker, R. J.: Antimicrob. Agents Chemother. *18*, 130 (1980)
86. Horng, C. B., McLimans, W.: Biotech. Bioeng. *17*, 713 (1975)
87. Clark, J. M., Gebb, C., Hirtenstein, M. D.: Develop. Biol. Stand. *50*, 81 (1982)
88. Clark, J. M., Gebb, C., Hirtenstein, M. D.: Eur. J. Cell Biol. *22*, 601 (1980)
89. Katinger, H. W., Scheirer, W., Kroemer, E.: Chem. Ing. Tech. *50*, 472 (1978)
90. Fleischaker, R. J., Sinskey, A. J.: Eur. J. Appl. Microbiol. *12*, 193 (1981)
91. Green, M., Henle, G., Deinhardt, F.: Virology *5*, 206 (1958)
92. Katinger, H., Scheirer, W.: Mass cultivation, production of animal cells, in: Animal Cell Biotechnology, vol. 1 (Spier, R. E., Griffiths, J. B., eds.) p. 167, Academic Press 1985
93. Clark, J. M., Hirtenstein, M. D.: Annals N.Y. Acad. Sci. *369*, 33 (1981)
94. Kilburn, D. G., Webb, F. C.: Biotech. Bioeng. *10*, 801 (1968)
95. Whiteside, J. P., Farmer, S., Spier, R. E.: Develop. Biol. Stand. *60*, 283 (1985)
96. Sinskey, A. J., Fleischaker, R. J., Tyo, M. A., Giard, D. J., Wang, D. I. C.: Annals N.Y. Acad. Sci, *369*, 47 (1981)
97. Boraston, R., Thompson, P. W., Garl, ,S., Birch, J. R.: Develop. Biol. Stand. *55*, 103 (1984)
98. Taylor, W. G., Richter, A., Evans, V. J., Sanford, K. K.: Exp. Cell Res. *86*, 152 (1974)
99. Taylor, W. G., Camalier, R. F., Sanford, K. K.: J. Cell Physiol. *95*, 33 (1978)
100. Balin, A. K., Goodman, D. B. P., Rasmussen, H., Cristofalo, V. J.: ibid. *89*, 235 (1976)
101. Radlett, P. J., Telling, R. C., Whiteside, J. P., Maskell, M. A.: Biotech. Bioeng. *14*, 437 (1972)
102. Kilburn, D. G., Lilly, M. D., Self, D. A., Webb, F. C.: J. Cell Sci. *4*, 25 (1969)
103. Taylor, G. W., Kondig, J. P., Nagle, S. C. Jr., Higuchi, K.: Appl. Microbiol. *21*, 928 (1971)
104. Young, D. V., Nakano, E. T.: Biochem, Biophys. Res. Commun. *93*, 1036 (1980)
105. Reuveny, S., Velez, D., Macmillan, J. D., Miller, L.: J. Immunol. Methods *86*, 53 (1986)
106. Tritsch, G. L., Moore, G. E.: Exp. Cell Res. *28*, 360 (1962)
107. Iio, M., Moriyama, A., Murakami, R.: Effects on cell proliferation of metabolites produced by cultured cells, their removal from culture in defined media, in: Proc. Int. Symp. on 'Growth, differentiation of cells in defined environment (Murakami, R., ed.) p. 437 Springer-Verlag 1985
108. Jensen, E. M., Lui, O. C.: Proc, Soc. Exp. Biol. & Med. *107*, 834 (1961)
109. McLimans, W. F., Blumenson, L. E., Repasky, E., Ito, M.: Cell Biol. Int. Reports *5*, 653 (1981)
110. Griffiths, J. B.: Cell products: An overview, in: Animal Cell Biotechnology, vol. 2 (Spier, R. E., Griffiths, J. P., eds.) p. 3, Academic Press 1985
111. Reuveny, S.: Adv. Biotech. Processes *2*, 1 (1983)
112. Giard, D. J., Thilly, W. G., Wang, D. I. C., Levine, D. W.: Appl. Environ. Microbiol. *34*, 668 (1977)
113. van Hemert, D., Kilburn, D. G., van Wezel, A. L.: Biotech. Bioeng. *11*, 875 (1969)
114. Meignier, B., Maugeot, H., Favre, H.: Develop. Biol. Stand. *46*, 249 (1980)
115. Mered, B., Albrecht, P., Hopps, H. E., Petricciani, J. C., Salk, J.: J. Biol. Stand. *9*, 137 (1981)
116. Tyo, M., Wang, D. I. C.: Adv. Biotech. *1*, 141 (1980)
117. Merck, W. A. M.: Develop. Biol. Stand. *50*, 137 (1981)
118. Giard, D. J., Loeb, D. H., Thilly, W. G.: Biotech. Bioeng. *21*, 433 (1979)
119. Clark, J. M., Hirtenstein, M. D.: J. Interferon Res. *1*, 391 (1981)
120. Havell, E. A., Viliek, J.: Antimicrob. Agents Chemother. *2*, 476 (1972)
121. Giard, D. J., Fleischaker, R. J., Sinskey, A. J.: J. Interferon Res. *2*, 471 (1982)
122. Damme, J. V., Billiau, A.: Method, Enzymol. *78*, 101 (1981)
123. Spier R. E.: Develop. Biol. Stand. *50*, 311 (1982)

Serum-Free Growth of Human Hepatoma Cells. A Review

Patrizia Bagnarelli and Massimo Clementi
Institute of Microbiology, University of Ancona, Italy

Human hepatoma-derived cell lines are able to grow in serum-free, chemically defined media in the presence or in the absence of different supplements, such as trace elements. The capability of these cells to grow in the absence of hormones and growth factors reflects the autocrine potentiality of these cell lines; moreover they produce and release attachment, spreading and growth factors. A serum-free growth of these cells may be regarded as an useful tool for different approaches. This paper summarizes the principal aspects of serum-free culture of human hepatoma cell lines and, principally, points to aspects that might be usefully investigated by using serum-free cultured liver-derived cells.

1 Introduction

Efforts have been made in the past few years to establish cell lines from liver with differentiated functions of their original cells; presently almost all cell lines derived from human liver are of cancer origin. These lines, some of which have been studied in detail, were derived from tumors of histologically defined hepatocellular carcinomas and hepatoblastomas.

Since human hepatoma is closely related to hepatitis B virus (HBV) infection [1], several cell lines derived from human hepatomas possess specific DNA sequences and produce hepatitis B surface antigen (HBsAg) [2–4]. In nature HBV seems to infect only man [5] and, at present, this virus has not yet been propagated in cell cultures; for this reason the study of virus multiplication at the molecular level has been greatly hampered. Moreover, very little is known concerning virus gene expression and its regulation. Because they constitute one of the cell culture systems producing viral antigens, some human hepatoma-derived cell lines have been widely studied in the last ten years. Although at present there is no information on the possible mechanism(s) of progression of chronic HBV infection to hepatocellular carcinoma, the availability of human hepatoma cell lines has permitted an approach to the study of HBV strategy within cancer cells.

Hepatoma cells may maintain in vitro several properties expressed by hepatic cells in vivo, on the other hand, a loss of differentiated functions may be observed in primary cultures of normal hepatocytes [6, 7], indicating that gene expression of hepatoma cells is quite stable under in vitro culture conditions. For these reasons they are a useful model for studying the control and modulation of both normal and transformed cell phenotype.

The majority of the human hepatoma/hepatoblastoma-derived cell lines have been established in the last ten years. Many of them, as previously observed, have been seen to maintain differentiated functions in vitro, such as synthesis of liver specific proteins [8] and the presence of membrane receptors for peptide hormones [9–11]. In addition, several properties of these cells have been related to the presence of integrated HBV-DNA sequences and to the transformed phenotype. More recently a serum-free growth of six human hepatoma and hepatoblastoma cell lines has been reported [12].

Proliferation of normal diploid cells is under the control of exogenous growth factors, while transformed cells may show a lack of growth factor requirements. This autonomous growth of transformed cells might be due to the expression of any of the controlling elements along the normal mitogenic pathway (the growth factor, the membrane receptor, the intracellular signal system which leads to the initiation of DNA synthesis and cell division). The expressed factor(s), which function(s) as transforming protein in malignant cells, may be encoded by oncogenes or its (their) expression may be under the control of oncogenes [13].

This review summarizes the experimental data on serum-free growth of human hepatoma-derived cell lines; we also focus on the principal efforts to culture liver cancer cells under chemically defined conditions.

2 Serum-free Cell Cultures

In the last ten years several papers have pointed to the capability of different cell lines to grow in serum-free, chemically defined media (for a review see Barnes and Sato [14]).

Hormones are likely to play a major role in the proliferation of cells in vitro and most serum-free synthetic media are supplemented with a complex mixture of hormones [15–18]. Nevertheless, trace elements, such as selenium, cadmium and lithium improve cell proliferation under serum-free conditions [12, 19–25] and sometimes are essential [26]. Defined amounts of insulin, hydrocortisone, transferrin, and selenium are generally included in formulations of serum-free media. Equally, hormone-like growth factors, a group of mitogenic polypeptides such as epidermal growth factor (EGF), fibroblast growth factor (FGF), multiplication stimulating activity (MSA), and platelet-derived growth factor (PDGF) have been used in several culture conditions [12, 15, 16, 27–30]. Moreover, specific attachment proteins such as cold insoluble globulin [31] or serum spreading factor [32] have been included in media formulations to improve adhesion of serum-free maintained cells; several authors used polylysine-coated culture dishes [33] or pretreatment of growth substrates with fibronectin with the same aim [34, 35].

However, the optimization of serum-free media is often time-consuming; since some supplements known to enhance growth response can act synergistically or additively with others, any possible activity has to be evaluated. Additionally, in some cases conditioned media have been used to improve growth under serum-free culture conditions [23, 28]. Nevertheless, a period of adaptation to serum-free growth is almost the rule and a longer lag phase has been observed in nearly all growth curves in the absence of serum.

3 Human Hepatoma-derived Cell Lines

Within the last ten years several cell lines have been established from primary human hepatoblastomas and hepatocellular carcinomas [36-38]; some of them showed interesting aspects, principally with regard to the maintenance of the biosynthetic capabilities of the normal liver parenchymal cells [8] (Table 1) and the production of HBsAg [39-42]. For these reasons they may be recognized as a major tool for investigation in several fields such as cell biology, virology, experimental endocrinology, experimental oncology, pharmacology and physiopathology of the liver.

Several tissue culture media have been used for growing hepatoma-derived cells; principally, minimal essential medium (MEM), William's medium and RPMI 1640,

Table 1. Plasma proteins produced and released by human hepatoma-derived cell lines

Plasma proteins	PLC/PRF/5	Hep3B	HepG2	HuH 1	HuH-4	HuH-7	HuH-6
Albumin	–/+[a]	+	+	+	+	+	+
Prealbumin	–	–	–	+	+	+	+
Transferrin	+	+	+	+	+	+	+
Fibronectin	+	+	+	+	+	+	+
Fibrinogen	+	+	+	+	+	+	+
Complement C_3'	+	+	+	+	+	+	+
Complement C_4'	–/+[a]	+	+	+	+	+	+
C_3' Activator	–	+	–	ND	ND	ND	ND
Plasminogen	+	+	+	ND	ND	ND	ND
α_1-Antitripsin	+	+	+	+	+	+	+
β_2-Macroglobulin	+	+	+	–	–	+	+
α_1-Antichymotripsin	–	+	+	ND	ND	ND	ND
α-Lipoprotein	–/+[a]	+	+	+	+	+	+
Haptoglobulin	–	+	+	–	+	+	+
Ceruloplasmin	+	+	+	+	+	+	+
α-Fetoprotein	+	+	+	+	–	+	+
Retinol-binding-protein	–	+	+	ND	ND	ND	ND

Supernatants from confluent cultures of mentioned hepatocarcinoma and hepatoblastoma cell lines in serum-free conditions were analysed by the Ouchterlony double-diffusion immunoprecipitation technique for detection of plasma proteins after 10-fold [8] or 500-fold [12] concentration.
ND = not done
[a] Discrepancies in results obtained by Knowles et al. [8] and Nakabayashi et al. [12].

supplemented with fetal calf serum (FCS), antibiotics and non-essential amino acids, have been employed by different researchers. However, the growth of these cell lines in media supplemented with FCS and other undefined substances did not allow a systematic study in vitro of the metabolism of these cancer cells under chemically controlled conditions; similarly, the identification of the effect of various agents and natural or synthetic drugs on cell growth, production of substances and HBV gene expression, has been greatly hempered. This suggests that a good model for the investigation of both oncogenesis and regulatory mechanisms of gene expression might be achieved by a serum-free, chemically defined culture medium for human hepatoma cells.

4 The PLC/PRF/5 Human Hepatoma Cell Line

In 1976, a continuous cell line derived from a primary liver cancer of man was shown to produce hepatitis B surface antigen (HBsAg) [37,43,44]. Because of its characteristic production and release of HBsAg in the culture medium, this cell line has been widely studied recently. Since PLC/PRF/5 cells may be regarded as the prototype of a set of hepatoma cell lines producing HBsAg and because they currently play an important role in the study of virus/host relationship, it may be of interest to describe the principal characteristics of these cells.

PLC/PRF/5 cells were derived from the liver of a Mozambican male whose serum was positive for HBsAg; the cells released HBsAg in culture medium in the form of both "22nm"-particles and filaments [41,45], but studies of the supernatant fluid and cell extracts failed to provide evidence of morphologically intact virions (Dane particles) or HBV infectivity [40,46]. In 1980, three reports from as many groups showed the presence of integrated HBV-DNA in the host genome [2-4]; these cells contain at least six (four complete and two partial) HBV genomes in integrated form; moreover the presence of RNA transcripts specific for the surface antigen sequences of HBV-DNA were detected, while no detectable transcripts corresponding to the hepatitis B core antigen were found [4]. More recently it was shown that the integrated DNA sequences of PLC/PRF/5 cells are methylated [47]. It was further demonstrated that viral DNA sequences coding for HBsAg are less (if at all) methylated than those coding for the major core polypeptide. Since DNA methylation may play a role in gene expression and, as a general rule, it appears that low levels of DNA methylation correlate with active transcription of genes, these data might account for the lack of expression of HBcAg-DNA sequences in PLC/PRF/5 cells.

Attempts to investigate the control of HBsAg synthesis in PLC/PRF/5 cells, as well as other human hepatoma cell lines such as Hep3B, failed to induce the production of Dane particles [48,49], while a recombinant plasmid (containing four copies of the HBV genome introduced into mouse and rat fibroblasts) and bone marrow cultures (obtained from a patient with acute HBV infection with an extrachromosomal form of HBV-DNA) were shown to produce Dane particles [50,51].

Different approaches have been used to study the dynamics of HBsAg production by both the parental PLC/PRF/5 cell line and clones which were established by plating procedures [42,52]. In addition, HBsAg has been detected by immunofluorescence in the cytoplasm and on the surface of PLC/PRF/5 cells [9]; the cells continuously release it into the culture medium, but the maximum rate of production is transitorily related to cytological changes preceding cell death [52]. More recently, in partial contrast with these observations, we showed that the maximum production of HBsAg by PLC/PRF/5 cells cultured under serum-free conditions is related to a reduction of the growth rate of viable cells [53].

Polyvalent anti-HBs antibodies do not affect the overall growth rate of these cells in vitro [54], however, in the presence of complement some monoclonal anti-HBs antibodies are able to induce cell death [55]. Moreover, in nude mice treated with monoclonals, tumor formation was significantly reduced [56].

These cells have also been widely used to better study in vitro the effect of natural and synthetic compounds on the expression of integrated HBV genes [57-60].

The antigen synthesized by two HBsAg producer cell lines (PLC/PRF/5 and Hep3B) has the same physical properties [8,61] and gives a polypeptide profile similar to that obtained by HBsAg derived from human serum [8,39,62]. The major polypeptide of HBsAg is glycosilated and glycoproteins, which are surface structures of enveloped viruses, can be involved in viral attachment [63], while in most cases the absence of these glycoproteins prevents virus budding. In order to evaluate the role of the carbohydrate portion of HBsAg on antigenicity and antigen expression, PLC/PRF/5 cells were cultured in the presence of inhibitors of glycosilation. It was shown that antigen release in a culture medium is not influenced by underglycosilation induced by several inhibitors [64,65].

Finally, the antigen produced by the human hepatoma cell line PLC/PRF/5 was purified by affinity chromatography by using monoclonal antibodies and applied for diagnostic purposes, such as Elisa techniques for anti-HBs detection in sera [66].

In conclusion, these cell lines and other human hepatoma cell lines which have been established and cultured in vitro represent a useful model; as more information on the biology of both HBV and human hepatocellular carcinoma cell lines becomes available, a clearer understanding of the basic events leading to this form of cancer of man should emerge.

5 Serum-free Growth of Human Hepatoma Cell Lines

The first evidence of serum-free growth of hepatoma and hepatoblastoma cell lines was reported by Nakabayashi et al. in 1982 [12]. The cell lines under study, reported in Table 2, were cultured under serum-free conditions using RPMI 1640 supplemented with selenium (as Na_2SeO_3) in order to study both growth and plasma proteins production. These authors also characterized the serum-free growth of the human hepatocellular carcinoma cell line named HuH-7; these cells were able to grow in chemically defined conditions even at low cell density, when 50% of a conditioned medium was added. Otherwise seeding input had to be enhanced. The HuH-7 cells grown in a fully synthetic medium retained differentiated functions of liver cells in vitro, with regard to both plasma proteins and liver enzymes production. Besides HuH-7 cells, all the cell lines which were studied were seen to be able to grow under serum-free conditions, with a single exception (the HLEC-1 cell line).

These authors emphasized that a period of adaptation to the new culture medium was not required for better growth when cells were plated and incubated in a condi-

Table 2. Human hepatoma cell line culture in a serum-free medium supplemented with selenium

Cell line	Cancer origin	Ref.
HLEC-1	HCC	67)
PLC/PRF/5	HCC	36)
HuH-1	HCC	68)
HuH-4	HCC	68)
HuH-7	HCC	12)
HuH-6	HB	69)

Human hepatocellular carcinoma (HCC)- and hepatoblastoma (HB)-derived cell lines cultured by Nakabayashi et al. [12] under serum-free, hormone-free conditions in RPMI 1640 supplemented with Na_2SeO_3

tioned medium of each cell line for the first two days. Moreover, a conditioned medium of the HuH-7 cell line was able to support the growth of HLEC-1 under serum-free conditions (at an intermediate seeding input). It follows that HuH-7 cells probably produce and release into the culture medium some hormones, growth factors or adhesion factors required for replication of HLEC-1 cells.

Finally, according to these researches, it may be postulated that all these hepatoma cell lines might be utilized for the production of a set of attachment or growth factors which are able to stimulate not only their own growth, but also that of other cells in serum-free conditions. More recently, we characterized the growth of PLC/PRF/5 cells under serum-free conditions [53]; we showed that the trace element selenium is not essential for cell replication and that an adaptation period is required for better growth of these cells.

Several growth parameters, such as attachment, doubling time and density saturation (Table 3) were observed to improve by comparing those shown by cells at the 6th and 23th passages in a serum-free medium. Moreover, when a conditioned medium (a 1:1 mixture of fresh RPMI 1640 and RPMI 1640 from 48-hour cultures of confluent

Table 3. Growth parameters of PLC/PRF/5 human hepatoma cells growing in the presence or in the absence of serum

Growth parameters	Culture conditions		
	10% FCS	10^{-8} M Na_2SeO_3 (6th p.)	10^{-8} M Na_2SeO_3 (23rd p.)
Seeding efficiency (%)	80	50	65
Doubling time (h)	31	43	36
Density saturation ($\times 10^3$ cm^{-2})	153	36	87

Comparison of several growth parameters between cells growing in the presence (RPMI 1640 plus 10% FCS) or in the absence of serum (RPMI 1640 plus selenium) and evidence of improved results after an adaptation period (from 6th to 23rd passages under serum-free conditions)

PLC/PRF/5 cells) was used, we were able to demonstrate a better attachment of cells but an almost similar growth rate.

The characterization of both attachment and spreading of the human hepatoblastoma cell line HepG2, cultured under serum-free conditions, was recently reported by Barnes and Reing [70]; this paper points to the synthesis, secretion and response of this cell line to the serum glycoprotein called human spreading factor (SF). This protein was seen to promote the attachment and spreading of several cells in culture and to influence cell growth [22, 71–73].

The HepG2 cell line can be grown in the absence of serum [74] and produces SF and a great many serum proteins under serum-free conditions, but although these cells produce both SF and fibronectin (a well-known attachment factor which is biochemically unrelated to SF) they are not able to respond (in terms of spreading) to fibronectin; on the other hand they showed a high spreading response to SF.

Moreover, examination of the effect of mitogens, such as polypeptide hormones and growth factors, used at concentrations known to be active in terms of proliferation on other cell lines, revealed a lack of activity by all these substances on both PLC/PRF/5 and HepG2 cell lines (with the exception of insulin for HepG2). Other supplementary factors tested for serum-free culture of both cell lines, such as prostaglandins F_2 and E_1, hydrocortisone, linoleic acid albumin, fibronectin, somatostatin and histydyl-lysine did not improve cell growth either [74, 75].

6 The Serum-free Approach for Hormone Receptors of Human Hepatoma Cells

We have used a serum-free approach to study the effect of insulin and the insulin binding level to the specific surface receptors of the PLC/PRF/5 human hepatoma cell line [76].

Since it had been noted that cell growth and phenotypic expression of the transformed cells had been shown to be functional and dependent on the various components present in the culture medium (in particular on the hormone component) [77], an influence of serum factors on the behavior of the specific binding and down-regulation of receptors was regarded as possible:

The PLC/PRF/5 hepatoma cell line was seen to possess specific membrane receptors for insulin with high affinity [10]; more recently we have observed that the absence of serum causes a rapid decrease of the number of PLC/PRF/5 specific insulin receptors and, later, the loss of down-regulation. In our opinion, these observations are consistent with the hypothesis of a general mechanism of adaptation of these tumor cells to different growth conditions; moreover the observed behavior of insulin receptors point to the role of serum factors for the maintenance of normal insulin binding and receptor down-regulation. This approach confirmed our opinion of the usefulness of serum-free culture of liver-derived tumor cells; it may indeed be used to study hepatocarcinogenesis in the absence of interfering substances.

7 Conclusions

Primary cancer of the liver is a common disease of man in regions where chronic carriers are common and much less common were they are not. In high incidence areas primary liver cancer is a disease of young adults [78-80]. Some hypotheses concerning the ethologic relationship of HBV to primary liver cancer of man might be usefully tested by using the in vitro models now available; among these, cell lines derived from human hepatoma tissue may be regarded as a major tool for scientists. During the last ten years these cell lines have supplied information on both the molecular biology of HBV and the study of the transformed phenotype. More recently the possibility of culturing cells under serum-free, hormone-free, chemically defined conditions has been of particular interest. Several differentiated properties expressed by these cells in culture (such as secretion of plasma proteins, enzyme activities, cell surface molecules and hormone receptors) may be investigated in the absence of interfering substances/activities of serum. Moreover, these cells express some properties due to their cancer origin.

It is presently necessary to develop studies in order to answer some major questions that still remain: Is HBV a co-carcinogen, rather than an oncogenic virus? Is the presence of integrated HBV-DNA sequences related to chromosomal rearrangements observed in these cells? Are there common phenotypic alterations in all HBV-related human hepatocellular carcinoma cell lines? Comparative data obtained by studying several hepatocarcinoma- and hepatoblastoma-derived cells under strictly controlled conditions are clearly necessary. All the aspects regarding the possibility of studying in vitro the differentiated functions of hepatocytes that these cancer cells have retained (and express in culture) also need the same controlled culture conditions.

In conclusion, the serum-free approach to hepatoma cells opens up several possibilities and it principally offers molecular virologists and cellular biologists a unique tool for the study of both HBV, a virus involved in the development of a human cancer, and hepatocyte functions.

8 Acknowledgements

This paper was partially supported by a grant (No. 830063152) of the Italian "Consiglio Nazionale delle Ricerche" (C.N.R.), "Progetto Finalizzato Controllo della Malattie da Infezione".
We wish to thank Dr. G. Carloni for helpful discussions.

9 References

1. Maupas, P., Melnick, J. L.: Prog. Med. Virol. *27*, 1 (1981)
2. Chakraborty, P. R., Ruiz-Opazo, N., Shouvald, D., Shafritz, D. A.: Nature (London) *286*, 531 (1980)
3. Bréchot, C., Pourcel, C., Louise, A., Rain, B., Tiollais, P.: Nature (London) *286*, 533 (1980)
4. Edman, J., Gray, P., Valenzuela, P., Rall, L. B., Rutter, W. J.: Nature (London) *286*, 535 (1980)
5. Zuckerman, A. J.: Persistence of hepatitis B virus in the population. In: Virus Persistence (eds. Mahy, B. W. J., Minson, A. C., Darby, G. K.), p. 39, Cambridge, University Press 1982

6. Tokiwa, T., Nakabayashi, H., Miyazaki, M., Sato, J.: In Vitro *15*, 393 (1979)
7. Auberger, P., Samson, M., Le Cam, A.: Biochem. J. *210*, 361 (1983)
8. Knowles, B. B., Hove, C. C., Aden, D. P.: Science *209*, 497 (1980)
9. Gerber, M. A., Garfinkel, E., Hirschman, S. Z., Thung, S. N., Panagiotatos, T.: J. Immunol. *126*, 1085 (1981)
10. Clementi, M., Testa, I., Bagnarelli, P., Festa, A., Pauri, P., Brugia, M., Calegari, L., De Martinis, C.: Arch. Virol. *81*, 177 (1984)
11. Clementi, M., Bagnarelli, P., Manzin, A., Testa, I., Festa, A.: Abstracts of the International Meeting on Advances in Virology, Catania (Italy), May 15–18, 1985
12. Nakabayashi, H., Takata, K., Miyano, K., Yamane, T., Sato, J.: Cancer Res. *42*, 3858 (1982)
13. Heldin, C. H., Westermark, B.: Cell *37*, 9 (1984)
14. Barnes, D., Sato, G.: Cell *22*, 649 (1980)
15. Bottestein, J., Hayashi, I., Hutchings, S., Masui, H., Mather, J., McClure, D. B., Ohasa, S., Rizzino, A., Sato, G., Serrero, G., Wolfe, R., Wu, R.: The growth of cells in serum-free hormonal-supplemented media. In: Methods in Enzymology (eds. Jakoby, W. B., Pastan, I. H.), p. 94, New York, Academic Press 1979
16. Hayashi, I., Sato, G.: Nature (London) *259*, 192 (1976)
17. Hayashi, I., Larner, R., Sato, G.: In Vitro *14*, 23 (1978)
18. Mather, J., Sato, J.: Exp. Cell. Res. *124*, 215 (1979)
19. Maciag, G. T., Kelly, B., Cerundolo, J., Ilsley, S., Kelley, P. R., Gaundreau, J., Forand, R.: Cell. Biol. Int. Rep. *4*, 43 (1980)
20. Murakami, H., Masui, H.: Proc. Natl. Acad. Sci. USA *77*, 3464 (1980)
21. Bottestein, J., Sato, G.: Proc. Natl. Acad. Sci. USA *76*, 514 (1979)
22. Barnes, D., Sato, G.: Nature (London) *281*, 388 (1979)
23. Rockwell, G. A., McClure, D., Sato, G.: J. Cell. Physiol. *103*, 323 (1980)
24. Medina, D., Oborn, C. J.: Cancer Res. *40*, 3982 (1980)
25. Tomooka, Y., Imagawa, W., Nandi, S., Bern, H. A.: J. Cell. Physiol. *117*, 290 (1983)
26. McKeehan, W. L., Hamilton, W. G., Ham, R. G.: Proc. Natl. Acad. Sci. USA *73*, 2023 (1976)
27. Hutchings, S. E., Sato, G.: Proc. Natl. Acad. Sci. USA *75*, 901 (1978)
28. Bradshaw, G. L., Sato, G. H., McClure, D. B., Dubes, G. R.: J. Cell. Physiol. *114*, 215 (1983)
29. Sallie, O. A., Kapadia, M., Mills, B., Daughaday, W. H.: Endocrinology *115*, 520 (1984)
30. Wolfe, R. A., McClure, D. B., Sato, G.: J. Cell. Biol. *87*, 434 (1980)
31. Hook, M., Rubin, K., Holdberg, A., Obrink, B., Vaheri, A.: Biochem. Biophys. Res. Commun. *79*, 727 (1977)
32. Barnes, D., Wolfe, R., Serrero, G., McClure, D., Sato, G.: J. Supramol. Struct. *14*, 47 (1980)
33. McKeehan, W. L., Ham, R. G.: J. Cell Biol. *71*, 727 (1976)
34. Orly, J., Sato, G.: Cell *17*, 295 (1979)
35. McClure, D. B.: Cell *32*, 999 (1983)
36. Prozesky, O. W., Brits, C., Grabow, W. O. K.: *In vitro* culture of cell lines from Australia antigen positive and negative hepatoma patients. In: Liver (eds. Saunders, S. J., Terblanche, J.), London, Pitman Medical 1973
37. Alexander, J. J., Bay, E. M., Geddes, E. W., Lecatsas, G.: S. Afr. Med. J. *50*, 2124 (1976)
38. Aden, D. P., Fogel, A., Plotkin, S., Damjanow, I., Knowles, B. B.: Nature (London) *282*, 615 (1979)
39. Marion, P. L., Salazar, F. H., Alexander, J. J., Robinson, W. S.: J. Virol. *32*, 796 (1979)
40. Skelly, J., Copeland, J. A., Howard, C. R., Zuckerman, A. J.: Nature (London) *282*, 617 (1979)
41. Stannard, L. M., Alexander, J. J.: Lancet *ii*, 713 (1976)
42. Wen, Y. M., Copeland, J. A., Mann, G. F., Howard, C. R., Zuckerman, A. J.: Arch. Virol. *68*, 157 (1981)
43. Alexander, J. J., Bey, E. M., Whitcutt, J. M., Gear, J. H. S.: S. Afr. J. Med. Sci. *41*, 89 (1976)
44. Macnab, G. M., Alexander, J. J., Lecatsas, G., Bey, E. M., Urbanowicz, J. M.: Br. J. Cancer *34*, 509 (1976)
45. Shibayama, T., Watanabe, T., Kojima, H., Yoshikawa, A., Watanabe, S., Kamimura, T., Suzuki, S., Ichida, F.: J. Med. Virol. *13*, 205 (1984)
46. Tabor, E., Copeland, J. A., Mann, G. F., Howard, C. R., Skelly, J., Snoy, P., Zuckerman, A. J., Gerety, R. J.: Intervirology *15*, 82 (1981)
47. Miller, R. H., Robinson, W. S.: Proc. Natl. Acad. Sci. USA *80*, 2543 (1983)

48. Daemer, R. J., Feinstone, S. M., Alexander, J. J., Tuly, J. G., London, W. T., Wong, D. C., Purcell, R. H.: Infect. Immun. *30*, 607 (1980)
49. Oefinger, P. E., Bronson, D. L., Dreesman, G. P.: J. Gen. Virol. *53*, 105 (1981)
50. Gough, N. M., Murray, K.: J. Mol. Biol. *162*, 43 (1982)
51. Elfassi, E., Romet-Lemonne, J. L., Essex, M., Franges-McLane, M.: Proc. Natl. Acad. Sci. USA *81*, 3526 (1984)
52. Copeland, J. A., Skelly, J., Mann, G. T., Howard, C. R., Zuckerman, A. J.: J. Med. Virol. *5*, *257* (1980)
53. Bagnarelli, P., Brugia, M., Manzin, A., Clementi, M.: La Ricerca Clin. Lab. *15*, 151 (1985)
54. Alexander, J. J., McElligott, S., Saunders, R.: S. Afr. Med. J. *54*, 973 (1978)
55. Shouval, D., Wands, J., Zurawski, V., Isselbacher, K., Shafritz, D.: Proc. Natl. Acad. Sci. USA *79* 650 (1982)
56. Shouval, D., Shafritz, D., Zurawski, V., Isselbacher, K., Wands, J.: Nature (London) *298*, 567 (1982)
57. Desmyter, J., De Groote, G., Ray, M. B., Bradburne, A. F., Desmet, V., De Somer, P., Alexander, J. J.: Prog. Med. Virol. *27*, 103 (1981)
58. Clementi, M., Bagnarelli, P., Pauri, P.: Arch. Virol. *75*, 137 (1983)
59. Marshall, J., Coulepis, A., Pringle, R., Dimitrakakis, M., Gust, I. D.: Acta Virol. *27*, 429 (1983)
60. Clementi, M., Bagnarelli, P., Pauri, P., Calegari, L.: J. Med. Virol. *13*, 117 (1984)
61. Alexander, J. J., Mac Nab, G., Saunders, R.: Perspect. Virol. *10*, 103 (1978)
62. Monjardino, J., Crawford, E.: Virology *96*, 652 (1979)
63. Paine, L., Norrby, E.: J. Virol. *34*, 142 (1980)
64. Alexander, J. J., Van Der Merwe, C., Saunders, R., McElligott, S., Desmyter, J.: Hepatology *2*, 92s (1982)
65. Clementi, M., Bagnarelli, P., Pauri, P., Brugia, M.: Arch. Virol. *78*, 109 (1983)
66. Merten, O. W., Reiter, S., Scheirer, W., Katinger, H.: Dev. Biol. Stand. *55*, 121 (1984)
67. Doi, I., Namba, M., Sato, J.: Gann *66*, 385 (1975)
68. Huh, N., Utakoji, T.: Gann *72*, 178 (1981)
69. Doi, I.: Gann *67*, 1 (1976)
70. Barnes, D. W., Reing, J.: J. Cell. Physiol. *125*, 207 (1985)
71. Barnes, D. W., Sato, G.: Anal. Biochem. *102*, 255 (1980)
72. Barnes, D. W., Silnutzer, J., See, G., Shaffer, M.: Proc. Natl. Acad. Sci. USA *80*, 1362 (1983)
73. Barnes, D. W., Foley, J., Shaffer, M., Sinultzer, J.: J. Clin. Endocrinol. Metab. *59*, 1019 (1984)
74. Hoshi, H., McKeeham, W. L.: In Vitro *21*, 125 (1985)
75. Clementi, M., Bagnarelli, P., Brugia, M., Pauri, P., Manzin, A., Calegari, L.: Abstracts Volume of the XV Meeting of the "European Tumor Virus Group" (ETVG), p. 83, Urbino, Italy 1984
76. Testa, I., Clementi, M., Festa, A., Bagnarelli, P., Brugia, M., De Martinis, C.: Dev. Biol. Stand. *60*, 81 (1985)
77. Mountjoy, K. J., Holdaway, I. M., Finlay, G. J.: Cancer Res. *43*, 4537 (1983)
78. Blumberg, B. S., Larouze, B., London, W. T., Werner, B., Hesser, J. E., Millman, I., Saimot, G., Payet, M.: Am. J. Pathol. *81*, 669 (1975)
79. Beasley, R. P., Lin, C. C., Hwan, L. Y., Chien, C. S.: Lancet *ii*, 1129 (1981)
80. Munoz, N., Linsell, A.: Epidemiology of primary liver cancer. In: Epidemiology of cancer of the digestive tract (eds. Correa, P., Haensrel, W.) p. 161, The Hague, Marinus Nijhoff 1982

Serum-Free Cultivation of Lymphoid Cells

Ulf Bjare
National Bacteriological Laboratory,
105 21 Stockholm, Sweden

1 Introduction

During the last decades, cultivation of animal cells has become increasingly important for studies of basic cellular functions and for the production of biological materials. For several reasons there has been an intensified search for serum substitutes in cultivation media. Defined media are of great advantage when investigating cellular processes affected by growth factors, cellular mediators and other specific agents as the background of possible effects caused by serum components can be diminished. Purification of biological materials produced in serum-free media is simpler than if serum is present and in the choice of serum substitutes, biological compatibility can be considered when products are used for specific aims. Many laboratories have developed defined media that will permit serum-free propagation of specific cell lines. None of these media will offer a general solution for serum-free cell cultivation [1,2,3,4].

Lymphoid cells are cells derived from the lymphatic system. These cells are available in larger quantities from live individuals or at autopsy and can be isolated from several different tissue sources: 1. peripheral blood, 2. lymph nodes, 3. spleen, 4. thymus, 5. bone marrow and 6. for macrophages from ascitic fluid [5].

Besides being a very convenient source of human cellular material, as cells can be obtained from an ordinary peripheral blood sample, these cells are simple to cultivate as they can be grown in suspension without the need of a microcarrier. This growth characteristic simplifies scale up as a culture can be extended by dilution with fresh medium without the need of trypsinization to release cells from solid support and the redistribution onto new carrier material.

The ease with which many lymphoid cells can be grown offers desirable advantages above many other cell sources for cultivation to produce biological materials [5,8,9,80,114].

2 General Methodology for Growth of Lymphoid Cells

Lymphoid cells can be cultivated in ordinary stationary flasks when growing volumes up to a few hundred milliliters per flask. For larger volumes the surface to volume ratio is insufficient in standard flasks and the culture needs stirring and might even require the addition of CO_2 and O_2 in the gas phase.

When growing volumes in the range of a few hundred milliliters up to a few liters, spinner flasks are convenient but results are highly variable depending upon the design of the flask. Some available spinner flasks have unsuitable stirrers so that cells sediment if not stirred vigorously and the growth of sensitive cells is harmed.

Large quantities of lymphoid cells can be propagated in roller bottles using simple equipment. Media volumes of 1/3–1/2 the total flask volume can be used giving good growth [8,9].

Many lymphoid cells can be grown in bioreactors, that offer the possibility to control pH, O_2, CO_2 and usually give a better control of stirring than in spinner flasks. Some lymphoid cells aggregate when growing. Intensive stirring will brake these aggregates which might lower the cell yields. Stirring has to be both gentle and efficient. Total cell mass thus depends upon the design of the bioreactor. Traditional bioreactors for bacterial or mold cultivation often have to be modified to give optimal yields with suspension cells.

3 Serum-Free Cultivation

Willmer in his book "Tissue Culture" 1935 quotes a number of papers published in the 1920s on attempts to make defined media [46]. True, success was not great, but some progress was however made and the foundation was laid for our present media, many of which were developed in the 1950s. A step-by-step addition of specific components by later workers has led to media that will permit serum-free cultivation of many cell lines. Still present media do not offer optimal growth conditions in most cases.

Katsuta and Takaoka [2] and Higuchi [1] in 1973 reviewed serum-free cell cultivation. Little progress had been made with lymphoid cells while the cultivation of other cell types in substituted media has been attempted with some recent success.

Pospíšil in 1966 showed that rabbit lymphoid cells could be short-term cultivated in a medium in which the high molecular weight components were replaced by Carbowax 20 M at a concentration of 0.2% [11].

In 1972 there appeared several reports on the culture of lymphoid cells in serum-free media. Vischer reported the culture of mouse lymphoid cells [12]. Horvat et al. demonstrated mitogenic activity and migration inhibitory activity in supernatants of serum-free human lymphocyte cultures stimulated with concanavalin A [13] and Brucher et al. described antibody synthesis and proliferation of lymphoid cells grown in a serum-free medium [14].

In 1972 Buhl et al. [15] reported that they had maintained a human leukemia cell line for over five months in a serum-free medium containing methyl cellulose, fetuin or albumin. Coutinho et al. in 1973 showed the effect of in vitro activation by T and B cell mitogens on proliferation and antibody synthesis of mouse lymphocytes in the absence of serum [16].

Apart from Buhl's paper most early reports on serum-free culture of lymphoid cells deal with short-term culture. In many cases data are not presented, showing that normal cells propagated but were rather incubated to make it possible to follow the formation of some cellular component or monitor cellular effects with only marginal growth. In many of these short-term cultures no serum substitutes were added to the serum-free media.

Muzik et al. 1982 reported that human long-term lymphoblastoid cell lines that had adapted to a defined, serum-free medium showed the same expression in the presence or absence of serum when tested for a panel of cell surface markers, including surface immunoglobulins, Ia antigens, Fc and complement receptors, and T and B erythrocyte rosettes. When analysing the Q-banded karyotypes from one cell line no alterations were observed during shift to serum-free conditions [10].

Transfer of lymphoid cells from a serum-enriched medium to a serum-free medium has been reported to be done step-by-step by successively decreasing the amount of serum. During this transfer period growth might be extremely slow. For some cell lines this lag period can last for several weeks. During this long lag phase mortality might rise to high levels [2,9].

3.1 Media Composition

Lymphoid cells can be cultivated in Dulbecco's modified Eagle's medium or in an enriched modification of this medium containing additional amino acids and vitamins according to Iscove and Melchers [7] or in the traditional RPMI 1640 medium of Moore et al. [18].

Even a mixture of Dulbecco's medium and Ham's F12 medium [3] in proportion 1:1 has been used [21,47] or a mixture of Iscove's medium and RPMI 1640 in proportion 1:1 [9].

A general and thorough review of media and growth requirements has been written by Ham and McKeehan [48].

3.2 Media Supplements

Serum plays a multifunctional role for cell growth. It contains crucial nutrients, offers a physical environment important for cells and holds transfer factors necessary for cellular functions. Serum-substitutes are selected to fulfil these functions, to substitute for nutrients needed, to offer physical protection and maintain factor transfer.

Since the 1950s a lot of investigations have been presented for non-lymphoid cell lines grown in serum-free culture showing the effects of many different agents such as albumin, insulin, glucocorticoid hormones, thyroxine, growth hormone and other pituitary hormones.

Albumin, transferrin and lipids can replace serum for maintaining LPS-stimulated murine B lymphocytes in culture. Cells grow and mature to IgM and IgG secretion even at dilutions as low as a single reactive B cell per culture [7)].

Na-selenit was included as an obligatory supplement in the enrichment of Dulbecco's modified Eagle's medium as modified by Iscove and Melchers [7)] and has been used in several other defined media [26,27)].

Zinc chloride was included in Higuchi's medium for monolayer cultures [1)]. Kristensen et al. found that zinc chloride had a clear enhancing effect in an enriched RPMI-1640 medium with L-alanine, albumin, transferrin and sodium selenite as additional supplements when investigating the transformation of lymphoblasts from mouse spleen and thymus [26)].

3.2.1 Albumin and Other Major Protein Supplements

Albumin has been used widely as one essential substitute for serum in defined media for several different cell types and is also a common constituent in serum-free media for lymphoid cells.

Bergman et al. 1967 showed, by incorporation of ^{3}H-thymidine into lymphocytes, that short-term substitution of serum by albumin gave an uptake in the same range as in the presence of autologous plasma [25)].

Iscove and Melchers stressed the role of albumin as a carrier of fatty acids in stabilizing the cell membrane. They prepared delipidated bovine serum albumin that was included when making a lipid suspension used as a supplement in their medium for short-term incubation of lymphocytes [7)].

Table 1 presents a selection of investigations performed in media partially substituted by albumin. Both normal and transformed lymphoid cells multiply in these media.

Table 1. A selection of investigations with albumin as one serum-substitute

Type of cells	References
Lymphocytes	7, 23, 25, 26, 27, 62, 73, 118, 130, 131, 132)
Splenocytes	103)
Lymphoblastoid	8, 9, 113)
Lymphoma	10, 76, 107)
Myeloma	111)

Darfler et al. grew murine T-lymphoma cells in a medium containing casein instead of albumin as a major protein supplement [19]. Later, casein was replaced by catalase in a medium also containing insulin, testosteron and transferrin with a total protein content of 50 micrograms ml^{-1}. The catalase was proposed to function to degrade H_2O_2 which was present in the medium that could support the growth of several lymphoid cell lines, including human and murine hybridomas [21]. Even Na_2SeO_3 has been proposed to protect against the highly cytotoxic H_2O_2 [21]. Casein has been a substitute for albumin in other defined media for different lymphoid cell types [9, 132].

Polet and Spieker-Polet showed that lymphocytes, once activated by Con A, lost their proliferative activity unless protected by Cohn fraction VI or beta-lactoglobulin. These protective proteins did not promote growth. Only serum albumin was shown to stimulate growth. Protein-free medium abolished growth [23]. Later Spieker-Polet and Polet demonstrated that beta-lactoglobulin as well as serum albumin served as carrier proteins for lipids essential for growth of lymphocytes [24]. Albumin was considered to be required to maintain the structure of lymphocyte membranes [23].

Yamamoto et al. used 0.1% lactalbumin hydrolysate as a serum substitute for different human B lymphoid cell lines [92].

No conclusive evidence has been published demonstrating an absolute need for a specific protein as a serum substitute. The presence of albumin in serum-free media is obviously not an absolute requirement for propagation of lymphocytes and other lymphoid cells. The major protein supplement is readily exchangeable and seems to function as a rather unspecific carrier for other essential components.

3.2.2 Transferrin

Hayashi and Sato showed that transferrin in combination with a series of hormones was essential for the growth of different non-lymphoid cell lines in the absence of serum [28].

The role of transferrin for growth of erythroid cells and its mechanism of delivery of iron to the cell was described by Morgan [29]. Its role for growth of lymphoid cells in serum-free media has, as already mentioned, been described by Iscove and Melchers [7]. Lymphoid cells, in general, have an absolute requirement for transferrin for long-term growth. Some malignant cell lines can however grow in the absence of transferrin [32].

Table 2 shows a selection of investigations with transferrin as one serum substitute in media of diverse compositions.

Table 2. A selection of investigations with transferrin as one serum-substitute

Type of cells	References
Lymphocytes	7, 26, 27, 31, 73, 86, 118, 130, 131)
Lymphocytes, binding, uptake	82, 133, 134)
Lymphoblastoid	8, 9)
Splenocyte	22)
T-lymphoma	19, 21, 22, 107, 132)
B-lymphoma	8, 9, 10, 22 , 76, 107)
Leukemic	45, 108)

Darfler et al. demonstrated good growth for a large number of transformed lymphoid cell lines in a defined medium with transferrin as one essential ingredient [21, 22].

Brock elucidated the role of transferrin in the transformation of mouse lymphocytes in a serum-free medium in response to concanavalin A. Transformation was dependent on the presence of homologous or heterologous transferrin. Optimal transformation was achieved at 10–50 micrograms ml^{-1} at 30–70% iron saturation. Transferrin was shown to bind to the transforming cells and was released undegraded after delivery of bound iron. Iron chelates did not promote transformation but were capable of delivering iron to lymphocytes. Thus, transferrin might fulfil functions additional to that of supplying iron [31].

Human lactoferrin was reported to be an essential growth-promoting factor and to have higher growth stimulatory activity than human transferrin in a medium developed for a human lymphocytic cell line. Long-term cultivation was achieved in a defined medium supplemented with human lactoferrin only. This medium also supported growth of various other human B- and T-lymphocytic cell lines. No stimulation of growth was achieved with mouse lymphocytic cell lines [30].

3.2.3 Insulin and Other Hormonal Supplements

Partial substitution of serum by addition of insulin in a serum-free medium was demonstrated by Lieberman and Ove in 1959 working with HeLa cells [17].

Brucher et al. 1972 reported that insulin and hydrocortisone were among several substitutes initially considered essential for the survival and proliferation of human and rabbit lymphoid cells but were subsequently eliminated as they were found non-essential. This was also demonstrated to be true for the initiation of the in vitro secondary immune response and the maintainance of stimulated antibody production [14].

Table 3 shows a selection of investigations with insulin as one serum substitute in composite media.

Table 3. A selection of investigations with insulin as one serum-substitute

Type of cells	References
Lymphocytes	[20, 118, 130]
Cytotoxic and helper T-cells	[131,]
T-lymphoma	[19, 21, 132]
B-lymphoma	[76]
Leukemic	[45]

Mendelsohn et al. reported that mitogen-stimulated human lymphocytes proliferated in albumin-free medium supplemented with insulin, transferrin, ethanolamine and selenium but without lipid addition. This medium even supported the growth of murine T-lymphocytes for up to two weeks [20].

A combination of testosteron and insulin were among a series of substitutes in a medium used by Darfler et al. when studying metabolic functions in T-cell lymphoma cells [19,21].

Brown et al. added insulin to the medium of Iscove and Melchers [7] for long-term growth of different lymphoid cells types [132]. Yen and Duigou [130] reported of two basal media supplemented with insulin, albumin and transferrin used for propagation of a human lymphocyte cell line and stimulated blood lymphocytes. A similar medium was used by Sharath et al. when studying immunoglobulin synthesis [118].

3.2.4 Lipids

The role of cholesterol for growth of cells in defined media has been long known [33,34,35]. Iscove and Melchers added soybean licithin and cholesterol to delipidated serum albumin. This mixture, was sonicated to prepare a suspension used for short-term culture of lymphocytes [7].

Cholesterol has been used in many different defined media for growth of lymphocytes [130] and in combination with l-alpha-phosphatidylcholine for growth of lymphoma cells [132] and leukemic cells [108].

Several workers have demonstrated that linoleic acid is essential for growth in an albumin-free medium for non-lvmphoid cells [42,43]. Linoleic acid was added to a serum-free medium for lymphoma cells [21] and was later used for many transformed lymphoid cell lines [22]. Linoleic acid was later substituted by dilinoleoyl phosphatidylcholine that also supported growth of other lymphoid cells, including human and murine hybridomas [19].

Spieker-Polet and Polet reported that optimal growth of stimulated lymphocytes was obtained with a combination of palmitic and oleic acids and a carrier protein. These fatty acids could be replaced by other saturated and unsaturated fatty acids [24].

Katsuta and Takaoka 1973 wrote an excellent summary of the state of the art of cultivation in protein-free and lipid-free synthetic media. A great deal of transformed cell lines have for a long time, been grown in simple media with different high molecular weight substitutes without lipid addition [2].

Uittenbogaart et al. [107] demonstrated that lipids were not essential for long-term growth of several different human T-cell and B-cell lines cultivated in the medium of Iscove and Melchers [7] with exclusion of lipids.

Many cell lines will adapt to poorer media during prolonged subcultivation. However, for newly established cell lines that are "normal", i.e. non-transformed, lipids are generally required for growth. Lymphoblastoid cells and lymphoma cells that can adapt to cultivation in serum-free medium will initially require lipid for growth [8,9]. During prolonged cultivation these cells will undergo secondary transformations and loose their lipid requirement.

Lipid preparations have to be standardized when used in cell cultivation media. Commercial preparations are available that are perfectly stable as suspensions when used in cultivation media. These liquid preparations have been developed for intravenous nutrition and are thus biologically compatible and well suited for production of material that will ultimately be used for clinical purposes [8,9,44].

3.2.5 Miscellaneous Supplements

Enhancement of antibody synthesis in vitro by beta-mercaptoethanol was reported by Click et al.[36] and has been included in the enriched Dulbecco's medium of Iscove and Melchers who as an alternative also used alpha-thioglycerol[7]. These additives were not required for growth but essential when studying immunoglobulin synthesis.

Ethanolamine was shown to be stimulatory in an albumin-free, serum-free medium for human lymphocyte proliferation[20] and for growth of hybridoma cells in serum-free medium[37].

The use of peptones as serum substitutes for mammalian cells in culture was reviewed by Taylor and Parshad 1977[40]. Proteose peptone was shown to stimulate proliferation of human lymphoma cells[38].

Disodium beta-glycerophosphate monohydrate was reported to have improved growth and viability of a murine T lymphoma cell line[19].

Kuchler observed that methyl cellulose protected cells in serum-free suspension cultures[39] and has also been used in cell mediator studies of lymphoid cells[83].

Fibronectin was one supplement beside albumin, transferrin and insulin used to study growth and maturation to IgM and IgG secretion of B lymphocytes[41].

Alfa-cyclodextrin was used as a partial substitute for bovine serum albumin in serum-free culture of a human lymphoblastoid cell line[49].

Many investigations have been performed in "conditioned media". These might in some cases by considered serum-free but are generally not well defined but useful when studying the effect of growth factors. These media will not be dealt with further in this article[20,40].

4 Serum-free Investigations

Many investigations listed below contain data that could equally well merit the respective papers to be quoted under one of the other sections but have been arbitrarily put into one of the following sections due to the main theme of each paper. Space will not permit a thorough discussion of results presented. The publications cited are representative examples of the types of studies listed and are not supposed to be a complete list of investigations made.

4.1 Immunological

Table 4 presents some immunological investigations performed in serum-free media of varying composition. Investigations dealing basically with lymphokines are presented in the next section. Many investigations have been performed in media that are literally "serum-free" and not substituted in any way.

Fathman and Fitch have given an excellent overview of long-term culture of immunocompetent cells[60]. Most of the investigations presented by them apply to serum-enriched media but many of the methods might also be used in substituted media.

Table 4. A selection of immunological investigations in serum-free media

Type of investigation	Reference
Immunoglobulin synthesis	116, 117, 125, 127, 129)
secretion, complement	7, 41, 66, 118)
macrophage interaction	65)
Polyclonal B-cell stimmulation, alpha-macroglobulin	63, 75, 119)
Immunosuppressive factors	125)
Immune response suppressor	123)
Chemiluminescence and immune cell activation	126)
IgE-binding, Ig-binding factors	128)
β-2-microglobulin	122)
Delayed-type hypersensitivity	124)

Hagiwara and Sato demonstrated that a human × human hybridoma produced significantly higher amounts of monoclonal antibody in serum-free medium than in serum-containing medium [62].

Leclercq et al. showed that stimulation of resting B cells to proliferation and differentiation into IgM-secreting cells takes place in serum-free medium by the addition of helper T-cell supernatant [63].
Immunoregulatory effects of oxidative metabolites of arachidonic acid on proliferation of B-lymphocytes were assessed in serum-free culture [61].

A large number of the investigations listed in Table 4 suggest that the results presented were affected by the absence of serum. Thus serum-free cultivation gives additional information that is distinct from that received in the presence of serum.

4.2 Lymphokines and Other Cellular Mediators

Table 5 presents a selection of investigations on cellular mediators made in serum-free medium. Lymphokine studies are also included in the above section on immunological investigations.

A general discussion of lymphokine regulation of T cell and B cell functions has been given by Smith [70].

Mononuclear cells of human peripheral blood could be cultured in serum-free medium for 1–2 months without loss of viability. Cells secreted prostaglandin and formed cytoplasmic extensions which created a dense cellular network which was not formed when the cells were cultured in serum-enriched media [71].

Soluble factors that induced the cytolytic activity and the expression of T cell growth factor receptors of a rat × mouse T cell hybrid were characterized from serum-free culture medium [95]. The human T-lymphoblast cell line, Mo, secretes a number of lymphokines, including an erythroid-potentiating activity which was purified and characterized from serum-free Mo-conditioned medium [96].

Asada et al. working with human T-cell hybridomas demonstrated four-fold larger amounts of lymphotoxin under serum-free conditions than under serum-containing conditions [98].

Table 5. Investigations on cellular mediators, lymphokines

Type of cells	References
Lymphocytes	52, 53, 55, 102, 135, 138)
From phagocytes	136)
From leukemic cells	137, 141)
Inhibitor of DNA-synthesis	139)
Migration inhibitory factor	82, 145)
Stimulation of granulocytic colony formation	83)
Chemotactic inhibitor	140)
Interleukin	52, 53, 106, 132, 145, 146, 147, 148)
T-cell derived growth factor	144)
T-cell lymphokine	141)
Mycoplasma induced lymphokine	142)
Cytolytic activity	143, 145)

Identification and characterization of many of the cellular mediators presented in Table 5 have been simplified by the absence of serum and results are not only quantitatively but even qualitatively different from investigations performed in serum-enriched media.

4.3 Interferon

Zoon et al. reported that Namalva cells could be cultured in serum-free medium to high cell densities and after viral induction gave interferon yields in the order of 10,000 units ml [114]. High interferon yields were reported in large scale culture of human lymphoma cells [9].

Imanishi et al. characterized an interferon from a human T leukemic cell line that was induced in serum-free medium [149]. Spontaneous interferon production and growth of lymphoblastoid cells in serum-free medium was reported by Sato et al. [113].

Tsujimoto et al. reported on production of human gamma-interferon in serum-free medium [150].

By growing the lymphoma cell line Daudi in serum-free medium containing insulin, transferrin and albumin it was demonstrated that growth inhibition caused by alfa-interferon could in part be due to a change in receptors for insulin [76].

4.4 Miscellaneous Studies

Table 6 presents a list of a heterogeneous mixture of investigations with lymphocytes in serum-free media of varying composition. Table 7 lists investigations for other lymphoid cells. There is no room for further discussion of these publications here.

Production of collagenase in mononuclear cells was demonstrated by Simpson et al. [151].

All three types of poliovirus replicated to good titers in lymphoblastoid cells isolated from patients with infectious mononucleosis. The propagated virus, injected into rabbits, gave antibody titers that complied with requirements for vaccine production [8].

Table 6. Investigations on lymphocytes in serum-free media

Type of investigation	References
Proliferation	27, 51, 54, 61, 99)
Mitogens	57, 69, 87)
Cytotoxins, lymphotoxins	58, 92, 93, 98)
Physiology	59)
Macrophage interaction	65)
Subpopulations	66)
Transformation	26)
Maturation	41)
Stim. of T-cells	68)
T-cell cytoxocity	72, 85)
T-cell factors	74, 79, 83)
T-cell, rheumatoid arthritis	81)
Suppression	73)
Polyclonal B-cell-stim.	63, 75, 119)
B-cell differentiation	78)
B-cell growth inhib.	88)
Migration inhib. factor	82)
Allotypes	84)
Iron deficiency	86)
Prostaglandins	67, 71, 97)
IgE synthesis	89)
Erythroid-potentiating activ.	90)
Chromosomal studies	91)
Role of pterins	94)

Table 7. Investigations on lymphoid cells in serum-free media

Type of cells	References
Splenocytes, thymocytes	26, 101, 103, 106)
Macrophages	51, 65, 74, 102)
Leukemic cells	100, 105, 107, 108)
Mast cells	104)
Lymphoma	9, 76, 107, 110, 113, 114)
Lymphoblastoid	8, 9)
Myeloma	50, 111)
Natural killer	112)

5 Large Scale Cultivation

Cultivation of lymphoid cells in bioreactors was described by Moore 1970 [5] and has been developed in many laboratories as a method to cultivate large quantities of cells and when producing interferon [77].

When the bioreactor volume is increased the cell density will usually decrease [77].

This is a common experience when growing different cell types in bioreactors and is due to the effects of scale up on the micro-environment.

Few reports have been published describing methodology for producing large quantities of lymphoid cells in serum-free media. Some of these were performed in bioreactors with serum-free media. There is no principal reason against using such media in bioreactor cultivation. A medium that gives good growth in 20-liter roller bottles [8,9] works equally well in a 100-liter bioreactor when growing lymphoblastoid cells to produce poliovirus and interferon.

A granulopoiesis inhibitor was partially purified from 200-liter serum-free cultures of porcine leukocytes grown in serum-free cultures [115].

6 Conclusions

Lymphoid cells offer a unique source of human cell material as samples can be taken in large amounts without harm to the donator. Even in animal studies this might be of crucial importance.

The possibility of growing these cells under serum-free, defined environmental conditions offers significant advantages before cultivation in serum-enriched media. Serum does not only stimulate growth, it even makes the interpretation of results more complicated as a complex, undefined mixture named serum is present.

As has been demonstrated in this article no universally accepted medium for serum-free culture of lymphoid cells has yet appeared. If one such medium should be nominated, it would presumably be the enriched Dulbecco's modified Eagle's medium as modified by Iscove and Melchers [7]. Many media are variations on this theme. Different cell lines have different requirements and the optimal medium for a specific cell line has to be worked out for each case.

In general, transformed cells have simpler requirements than more "normal" variants.

Measuring a range of biochemical parameters and cellular markers it has been demonstrated that no fundamental cellular changes are induced when transferring cells from a serum-enriched medium to a serum-free medium [10,120,121].

7 References

1. Higuchi, K.: Adv. Appl. Microbiol. *16*, 111 (1973)
2. Katsuta, H., Takaoka, T.: Methods Cell Biol. *6*, 1 (1973)
3. Ham, R. G. in: Growth of Cells in Hormonally Defined Media (Sato, G., Pardee, A., Sirbasku, D. eds.) p. 39, Cold Spring Harbor Laboratory 1982
4. Sato, G.: Nature *259*, 132 (1976)
5. Moore, G. in: Methods in Cancer Res. (Busch, H. ed.) *5*, 423 (1970)
6. Guilbert, L. J., Iscove, N. N.: Nature *263*, 594 (1976)
7. Iscove, N. N., Melchers, F.: J. Exp. Med. *147*, 923 (1978)

8. Bjare, U.: J. Virol. Methods *9*, 259 (1984)
9. Bjare, U., Räbb, I.: Develop. Biol. Standard. *60*, 349 (1985)
10. Muzik, H., et al.: In Vitro *18*, 515 (1982)
11. Pospíšil, M.: Folia Microbiol. *12*, 367 (1967)
12. Vischer, T. L.: J. Immunol. Methods *1*, 190 (1972)
13. Horvat, M. et al.: Int. Arch. Allergy Appl. Immunol. *43*, 446 (1972)
14. Brucher, J. et al.: Arch. Roum. Pathol. Exp. Microbiol. *31*, 105 (1972)
15. Buhl, S. et al.: Proc. Soc. Exp. Biol. Med. *140*, 1224 (1972)
16. Coutinho, A. et al.: Eur. J. Immunol. *3*, 299 (1973)
17. Lieberman, I., Ove, P.: J. Biol. Chem. *234*, 2754 (1959)
18. Moore, G. E., Gerner, R. E., Addison Franklin, H.: JAMA *199*, 519 (1967)
19. Darfler, F. J., Insel, P. A.: J. Cell Physiol. *115*, 31 (1983)
20. Mendelsohn, J., Caviles Jr., J., Casagnola, J. in: Growth of Cells in Hormonally Defined Media (Sato, G., Pardee, A., Sirbasku, D. eds.) p. 677, Cold Spring Harbor Laboratory 1982
21. Darfler, F. J., Murakami, H., Insel, P. A.: Proc. Natl. Acad. Sci. U.S.A. *77*, 5993 (1980)
22. Darfler, F. J., Insel, P. A. in: Growth of Cells in Hormonally Defined Media (Sato, G., Pardee, A. and Sirbasku, D. eds.) p. 717, Cold Spring Harbor Laboratory 1982
23. Polet, H., Spieker-Polet, H.: J. Immunol. *117*, 1275 (1976)
24. Spieker-Polet, H., Polet, H.: ibid. *126*, 949 (1981)
25. Bergman, B. et al.: Scand. J. Haemat. *4*, 176 (1967)
26. Kristensen, F. et al.: Scand J. Immunol. *16*, 209 (1982)
27. Kristensen, F. et al.: Immunol. Letters *5*, 59 (1982)
28. Hayashi, I., Sato, H. G.: Nature *259*, 132 (1976)
29. Morgan, E. H.: Molec. Aspects. Med. *4*, 3 (1981)
30. Hashizume, S. et al.: Bioch. Biophys. Acta, *763*, 377 (1983)
31. Brock, J. H.: Haematologia *17*, 187 (1984)
32. Hatzfeld, J., et al. in: Growth of Cells in Hormonally Defined Media (Sato, G., Pardee, A., Sirbasku, D. eds.) p. 703, Cold Spring Harbor Laboratory 1982
33. Sato, G., Fischer, H. W., Puck, T. T.: Science *126*, 961 (1957)
34. Higuchi, K.: In Vitro *6*, 239 (1970)
35. Higuchi, K., Robinson, R. C.: ibid. *9*, 114 (1973)
36. Click, R. E., Benck, L., Alter, B. J.: Cell. Immunol. *3*, 156 (1972)
37. Murakami, H., et al.: Proc. Natl. Aca. Sci. *79*, 1158 (1982)
38. Lazar, A. et al.: Dev. Biol. Stand. *50*, 167 (1982)
39. Kuchler, R. J., Marlowe, M. L., Merchant, D. J.: Exp. Cell Res. *20*, 428 (1960)
40. Taylor, W. G. et al.: Meth. Cell. Biol. (Prescott, D. ed.) *15*, 421 (1977)
41. Tanno, Y., Arai, S., Takishima, T.: J. Immunol. Meth. *52*, 255 (1982)
42. Ham, R. G.: Science *140*, 802 (1963)
43. Dubin, I. N., Czernobilsky, B., Herbst, B.: J. Nat. Cancer Inst. *34*, 43 (1965)
44. Carlson, L. A.: Scand. J. Clin. Lab. Invest. *40*, 139 (1980)
45. Breitman, T. R., Keene, B. R. in: Growth of Cells in Hormonally Defined Media (Sato, G., Pardee, A., Sirbasku, D. eds.) p. 691, Cold Spring Harbor Laboratory 1982
46. Willmer, E. N.: Tissue Culture, London, Methuen 1935
47. Barnes, D., Sato, G.: Cell *22*, 649 (1980)
48. Ham, R. G., McKeehan, W. L.: Methods in Enzymology, Vol. LVIII, p. 44, Academic Press, New York 1979
49. Yamane, I. et al. in: Growth of Cells in Hormonally Defined Media (Sato, G., Pardee, A., Sirbasku, D. eds.) p. 87, Cold Spring Harbor Laboratory 1982
50. Kawamoto, T. et al.: Analyt. Biochem. *130*, 445 (1983)
51. Liu, M. C. et al.: J. Immunol. *132*, 2895 (1984)
52. Schmitt, S., Schenkein, H. A.: J. Immunol. Methods *63*, 337 (1983)
53. Prystowsky, M. B. et al.: Fed. Proc. *42*, 2757 (1983)
54. Lipsky, P. E., Rosenthal, A. S.: J. Immunol. *117*, 1594 (1976)
55. Friedrich, W. et al.: Int. Arch. Allergy Appl. Immunol. *49*, 504 (1975)
56. Pauly, J. L., Han, T.: J. Lab. Clin. Med. *88*, 864 (1976)
57. Voitenok, N. N.: Tsitologiia *18*, 356 (1976)
58. Lee, S. C., Lucas, Z. J.: J. Immunol. *117*, 283 (1976)

59. Dumont, F.: Ann. Immunol. (Paris) *126C*, 453 (1975)
60. Fathman, C. G., Fitch, F. W. in: Fundamental Immunology, (Paul, W. E. ed.) p. 781 New York, Raven Press 1984
61. Goodmân, M., Weigle, W. O.: J. Allergy Clin. Immunol. *74*, 418 (1984)
62. Hagiwara, H., Sato, G. H.: Mol. Biol. Med. *1*, 245 (1983)
63. Leclercq, L., Bismuth, G., Thèze, J.: Proc. Natl. Acad. Sci. *81*, 6491 (1984)
64. Ishizaka, S.: Immunol. Lett. *6*, 343 (1983)
65. Lemke, H. et al.: Scand. J. Immunol. *4*, 707 (1975)
66. Melchers, F., Von Boehmer, H., Phillips, R. A.: Transplant. Rev. *25*, 26 (1975)
67. Okazaki, T. et al.: Prostaglandins *15*, 423 (1978)
68. Bick, P. H., Johnson, A. G.: Scand. J. Immunol. *6*, 1133 (1977)
69. Banck, G.: J. Immunol. Methods *51*, 279 (1982)
70. Smith, K. A. in: Fundamental Immunology, (Paul, W. E., ed.) p. 559, New York, Raven Press 1984
71. Halpern, J. et al.: J. Med. *15*, 1 (1984)
72. Spits, H. et al.: J. Immunol. *128*, 95 (1982)
73. Farrant, J., Newton, C.: Clin. Exp. Immunol. *45*, 504 (1981)
74. Frölich, M.: J. Cell Physiol. *109*, 439 (1981)
75. Chang, J. L. et al.: Immunology *44*, 745 (1981)
76. Faltynek, C. R., McCandless, S., Baglioni, C.: J. Cell Physiol: *121*, 437 (1984)
77. Finter, N. B., Fantes, K. H.: The purity and safety of interferons prepared for clinical use, in: Interferon, Vol. 2 (Gresser, I., ed.) p. 65, London, Academic Press 1980
78. Bich-Thuy, L. T. et al.: J. Immunol. *127*, 1299 (1981)
79. Frey, J. R., Lefkovits, I.: Exp. Cell Biol. *49*, 125 (1981)
80. Keay, L.: Methods Cell Biol. *20*, 169 (1978)
81. Slavin, S., Strober, S.: Ann. Rheum. Dis. *40*, 60 (1981)
82. Habasha, F. G. et al.: Am. J. Vet. Res. *46*, 1415 (1985)
83. Niskanen, E., Rahman, R.: Cancer Res. *45*, 3493 (1985)
84. Wolf, B., Yarmush, M. L.: Immunology *55*, 65 (1985)
85. Yssel, H., Spits, H., de Vries, J. E.: J. Exp. Med. *160*, 239 (1984)
86. Mainou-Fowler, T., Brock, J. H.: Immunology *54*, 325 (1985)
87. Low, C. E. et al.: J. Lipid Res. *25*, 1090 (1984)
88. Kawano, M., Iwato, K., Kuramoto, A.: J. Immunol *134*, 375 (1985)
89. Sarfati, M. et al.: Immunology *53*, 783 (1984)
90. Westbrook, C. E. et al.: J. Biol. Chem. *259*, 9992 (1984)
91. Nordenskjöld, M., Lambert, B.: J. Med. Genet. *21*, 129 (1984)
92. Yamamoto, R. S. et al.: Nature *308*, 641 (1984)
93. Aggarwal, B. B., Moffat, B., Harkin, R. N.: J. Biol. Chem. *259*, 686 (1984)
94. Ziegler, I., Hamm, U., Berndt, J.: Cancer Res. *43*, 5356 (1983)
95. Erard, F. et al.: J. Exp. Med. *160*, 584 (1984)
96. Westbrook, C. A. et al.: J. Biol. Chem. *259*, 9992 (1984)
97. Rutherford, B., Schenkein, H. A.: J. Immunol. *130*, 874 (1983)
98. Asada, M. et al.: Cell Immunol. *77*, 150 (1983)
99. Streilein, J. S., Hart, D. A.: Infect. Immun. *14*, 463 (1976)
100. Okano, H., Nishiyama, T.: Biomedicine *25*, 321 (1976)
101. Draber, P., Viklick, Y. V., Lengerova, A.: Folia Biol. (Praha) *23*, 305 (1977)
102. Shirahata, T. et al.: Z. Parasitenkd. *53*, 31 (1977)
103. Tamura, S. et al.: Microbiol. Immunol. *26*, 1065 (1982)
104. Ginsburg, H. et al.: Int. Arch. Allergy Appl. Immunol. *66*, 447 (1981)
105. Moolten, F. L., Schreiber, B.: J. Immunol. Methods *36*, 359 (1980)
106. Valyakina, T. I. et al.: Mol. Immunol. *21*, 811 (1984)
107. Uittenbogaart, C. H., Cantor, Y., Fahey, J. L.: In Vitro *19*, 67 (1983)
108. Taketazu, F. et al.: Cancer Res. *44*, 531 (1984)
109. Goldenberg, G. J., Froese, E. K.: Biochem. Pharmacol. *34*, 771 (1985)
110. Ling, M. et al.: J. Immunol. *134*, 1449 (1985)
111. Shrestha, K., Hiramoto, R. N., Ghanta, V. K.: Cell Immunol. *90*, 24 (1985)
112. Seymour, G. J. et al.: J. Periodontol. *55*, 289 (1984)

113. Sato, T. et al.: Exp. Cell Res.: *138*, 127 (1982)
114. Zoon, K. C., Bridgen, P. J., Smith, M. E.: J. Gen. Virol. *44*, 227 (1979)
115. Kastner, M. et al.: Z. Naturforsch. *39*, 639 (1984)
116. Brenner, M. K. et al.: Immunology *50*, 377 (1983)
117. Rastogi, S. C., Birkeboek, N. H.: Eur. Neurol. *22*, 131 (1983)
118. Sharath, M. D., Rinderknecht, S. B., Weiler, J. M.: J. Lab. Clin. Med. *103*, 739 (1984)
119. Rastogi, S. C., Clausen, J.: Immunobiologi *169*, 37 (1985)
120. Lundblad, G. et al.: Int. J. Biochem. *15*, 835 (1983)
121. Bjare, U., Lundblad, G., Räbb, I.: ibid. *17*, 67 (1985)
122. Kefford, R. F. et al.: Nature *308*, 641 (1984)
123. Aune, T. M., Webb, D. R., Pierce, C. W.: J. Immunol. *131*, 2848 (1983)
124. Tamura, S. I. et al.: Cell Immunol. *76*, 156 (1983)
125. Waldmann, H., Lachmann, P. J.: Eur. J. Immunol. *5*, 185 (1975)
126. Nakamura, M., Ishida, N., Kamo, I.: JNCI *65*, 759 (1980)
127. Tees, R., Schreier, M. H.: Nature *283*, 780 (1980)
128. Jensen, J. R., Sand, T. T., Spiegelberg, H. L.: Immunology *53*, 1 (1984)
129. Weiler, J. M. et al.: ibid. *46*, 247 (1982)
130. Yen, A., Duigou, R.: Immunol. Lett. *6*, 169 (1983)
131. Yssel, H. et al.: J. Immunol. Methods *72*, 219 (1984)
132. Brown, R. L. et al.: J. Cell Physiol *115*, 191 (1983)
133. Brock, J. H., Rankin, M. C.: Immunology *43*, 393 (1981)
134. Brock, J. H.: ibid. *43*, 387 (1981)
135. Persson, U., Möller, G.: Scand. J. Immunol. *4*, 527 (1975)
136. Rutherford, B., Steffin, K., Sexton, J.: J. Reticuloendothel. Soc. *31*, 281 (1982)
137. Olsson, I., Olofsson, T., Mauritzon, N.: JNCI *67*, 1225 (1981)
138. Warren, H. S., Pembrey, R. G.: J. Immunol. Methods *41*, 9 (1981)
139. Beneke, J. S. et al.: J. Immunol. *124*, 2950 (1980)
140. Donabedian, H.: J. Infect. Dis. *152*, 24 (1985)
141. Leung, K., Chiao, J. W.: Proc. Natl. Acad. Sci. USA *82*, 1209 (1985)
142. Proust, J. J., Buchholz, M. A., Nordin, A. A.: J. Immunol. *134*, 390 (1985)
143. Erard, F. et al.: J. Exp. Med. *160*, 584 (1984)
144. Benveniste, E. N. et al.: Proc. Natl. Acad. Sci. USA *82*, 3930 (1985)
145. Bödeker, B. G. et al.: Immunobiology *166*, 12 (1984)
146. Shipley, G. D. et al.: Cell Immunol. *93*, 459 (1985)
147. Ulrich, F.: ibid. *80*, 241 (1983)
148. Krakauer, T., Oppenheim, J. J.: ibid. *80*, 223 (1983)
149. Imanishi, J. et al.: Biken J. *25*, 79 (1982)
150. Tsujimoto, M. et al.: Infect. Immun. *41*, 181 (1983)
151. Simpson, J. W. et al.: J. Dent. Res. *59*, 2 (1980)

Rabies Vaccine Production in Animal Cell Cultures

Pierre Sureau
Head, Rabies Unit, Institut Pasteur, 28, rue du Dr. Roux,
75724 Paris Cédex 15/France

Vaccination against rabies was discovered by L. Pasteur a hundred years ago. If the original "Pasteur's treatment" is no longer used, various types of neural tissue vaccines, prepared from the brain of experimentally infected animals, are still produced and applied in many countries. These vaccines are poorly protective and often responsible for untoward reactions.

Cell cultures have introduced the rabies vaccines into a new and modern era.

From inoculated cell cultures rabies virus may be obtained in a pure form and in large amounts. Concentrated and purified virus may be used for the preparation of highly immunogenic and well tolerated vaccines, applicable either to pre-exposure immunization or to post-exposure rabies treatment.

With the newly introduced biotechnologies, such as the large bioreactors and the micro-carriers, these cell culture rabies vaccines may be produced at an industrial scale and at a reasonable cost.

1 Introduction

Rabies is a lethal virus disease of animals, which can be transmitted to man, a viral zoonose, which is known and feared since the greek and roman times. During more than twenty-five centuries it has been observed in most countries in the world in Europe, Asia, Africa and the Americas [1].

1.1 The Present Situation of Rabies in the World

Rabies occurs under two epidemiological aspects: the urban rabies where stray dogs are the reservoir and vector of the disease and the selvatic rabies where wild animals are involved.

Urban rabies, with dogs as the main source of human contamination, is still prevalent to-day in cities and villages of most countries in Asia, Africa and Latin America. Every year hundred of thousand of persons must receive the rabies vaccination after being bitten by rapid dogs and several thousand of fatal human cases are recorded.

On the contrary, and following the implementation of strict measures of sanitary and medical prophylaxis, urban rabies has been eradicated from Japan in Asia, from most of the European countries such as Iceland, the British Islands, the Scandinavian countries, Spain, Portugal, Greece and Bulgaria. Australia and New-Zeeland, as well as the Pacific islands are also rabies-free.

Selvatic rabies may be observed, in some countries, in parallel with urban rabies; the main wild animal species involved are the wolf in the Soviet Union, Central Asia, Iran; the jackal in Africa north of the Sahara; the mongoose in south Africa; the vampire bats in intertropical America.

But on the other hand, wildlife rabies has become presently the most important form of the disease in the industrial countries which had succeeded in eradicating urban rabies. In Europe for instance an epizootic of fox rabies which started from Poland around 1940 has progressively invaded, during the last forty years, East and West Germany, Czechoslovakia, Hungary, Austria, Switzerland, the south of Belgium, the east of France, the north of Italy and Yugoslavia. In the United States and Canada, wildlife rabies is also the only form of the disease presently observed; several animal species are involved according to the geographical areas: skunks, coyotes, raccoons, polar and red foxes, insectivorous bats. With wildlife rabies the transmission to man may occur either directly or through the contamination of domestic animals. Fortunately such human cases are very rare [2].

1.2 The Need for Rabies Vaccines

Considering this situation there is a persistent and important need for rabies vaccines to be applied for the control of animal rabies and for the protection of human beings exposed to the risk of contamination from rabid animals.

Rabies vaccines for veterinary use are applied mainly to domestic carnivores, dogs and cats, which constitute the most frequent source of human exposure. Such a vaccination is compulsory in some countries, only recommended in others. In many instances mass vaccination campaigns of dogs, associated to the elimination of stray dogs, have been very successful in controlling the disease. Vaccination of domestic herbivores (cattle, horses) is highly recommended for the protection of human health but has also economical reasons such as in Latin America where losses of cattle due to vampire bat transmitted rabies are important.

Rabies vaccination of the wildlife animal species has been successfully applied, on an experimental basis, in some European countries such as Switzerland and West-Germany to control the fox rabies enzootic.

Where ever animal rabies does exist and where human beings are exposed to rabies infection from rabid animals (through bite, scratch or contact) human post-exposure vaccination must be applied. Large amounts of rabies vaccine for human use are then requested. These vaccines should be both potent and well tolerated. The rabies immunization may also be applied as a preventive measure to people professionally exposed to rabies infection such as veterinarians, forest keepers and so on.

1.3 The Use of Animal Cell Cultures for Rabies Vaccines Production

After the discovery of Pasteur a century ago, rabies vaccines have been produced with rabies virus obtained from the brain of experimentally inoculated animals. Such vaccines are still produced and used in many developing countries. Unfortunately they are not always very immunogenic, and free of untoward reactions. Furthermore, it is difficult, using animal brains as a source of virus, to achieve large scale production.

The introduction of the animal cell culture technology in the field of rabies vaccines production about twenty-five years ago has opened a new era. Preparation of purified, concentrated, potent and well tolerated rabies vaccines has become feasible, for both medical and veterinary use. The introduction of the micro-carriers and bioreactors technology has recently made possible the production of these new vaccines in an industrial scale. It must be noted, nevertheless, that such modern technologies require sophisticated and costly equipment as well as high level trained personnel and that it may be difficult to transfer these technologies to the developing countries which are eager to use them in order to increase and improve their vaccine production. The World Health Organization is paying much attention and devoting much effort in trying to help resolving this situation.

2 History

It must be acknowledged that the first experimental immunization against rabies was achieved in sheep by means of intravenous inoculation of rabies virus by Galtier [3] in 1881. Nevertheless rabies vaccination of animals and man was actually invented by L. Pasteur.

2.1 Pasteur's Discovery, a Century ago

In 1881 Pasteur [4] in collaboration with Chamberland, Roux and Thuillier demonstrated the presence of the rabies virus in the central nervous system of rabid animals. By means of serial intra-cerebral passages in rabbits he obtained what he called the "fixed virus" which, contrary to the non adapted virus (called "street virus") was inducing in the experimental animals a regularly reproducible paralytic form of rabies after a short and constant incubation period [5].

He thought that the rabies virus could be used as a vaccine if its virulence could be attenuated, in a similar way he had been successfully using to develop vaccines against fowl cholera and anthrax disease. Pasteur achieved the progressive attenuation of virulence of the fixed virus by dessicating spinal cords of infected rabbits in dry air at room temperature. Then he developed a method for the immunization of dogs consisting in serial daily inoculation of spinal cord suspensions of progressively increasing virulence: all vaccinated dogs were "refractory" to a subsequent challenge with virulent virus inoculated by intra-cerebral route. This was a pre-exposure immunization but Pasteur considered also that, due to the long incubation period of rabies after the bite of a rabid animal, it could also be possible to induce this "refractory state" even if the vaccination was applied after the exposure, and that his method could be used for the treatment of bitten persons [6]. Soon after, on the 6th of July 1985, the Pasteur's treatment was applied, for the first time, to the young boy Joseph Meister, severely bitten 60 h previously by a rabid dog. Meister received, in 11 days, 13 injections of spinal cord suspensions of progressively increasing virulence. He was protected and survived [7]. After him hundred other contaminated people received the same tretment and were saved.

2.2 Rabies Vaccines from Pasteur's Time up to the Use of Animal Cell Cultures

One of the practical difficulties with the original Pasteur's method was to have enough available amount of spinal cords, at the various required degrees of attenuation, to simultaneously perform several treatments, started at different times. The solution was found by Calmette who demonstrated that attenuated cords could be preserved in glycerol at ice temperature, where their virulence was maintained for several days. This modified Pasteur's method has been used at the Pasteur Institute in Paris from 1912 to 1952.

Soon after Pasteur's discovery other methods were proposed to achieve the attenuation of the virus: Högyes [8] in 1887 used a dilution technique; Puscariu [9] in 1895 heated the virus suspensions for 10 min at decreasing temperatures when Babès [10] in 1912 applied heating at 58 °C for decreasing periods of time. All the above mentioned methods were using physical agents in order to obtain a progressive attenuation of the virulence. In 1908, Fermi [11] introduced the use of a chemical substance, phenol, for the attenuation of the virus; the Fermi type of rabies vaccine is still to-day produced in several countries; nevertheless its use is not recommended because of its residual virulence. Semple [12] in 1911 prepared and used in India a fully inactivated rabies vaccine, treated with phenol; the Semple type of vaccine, most often prepared from sheep or goat brain infected with Pasteur fixed rabies virus strain has been used for more than half a century in most countries in the world; it is still produced in several countries in Asia, Africa and even in Europe. A modification of the Semple's method using ether in association with phenol has been introduced in 1925 by Hempt [13]; this type of vaccine has been used until a few years ago in Yugoslavia. More recently Lépine [14] introduced the use of beta-propiolactone for the inactivation of the virus in the Semple's type rabies vaccine, in association with a reduced concentration of phenol.

All these neural tissue vaccines, which contain aside with the immunogenic viral antigen a non negligible amount of animal brain substances may cause, in the vaccinees, neuro-paralytic accidents due to a sensitization against the myelin content of the vaccine. In order to avoid this inconvenience, several investigators have proposed to use vaccines prepared from the brain of newborn rodents of less than 9 d of age at the time of harvest. The most widely used of this type of vaccines is the suckling mouse brain vaccine of Fuenzalida and Palacios [15]. This vaccine, inactivated with UV light, and freeze-dried, has replaced the Semple vaccine in all Latin-American countries and in several countries in Africa. Other vaccines of the same type are the new-born rat brain vaccine of Svet-Moldavskij [16] and the suckling rabbit brain vaccine of Gispen [17].

A different approach for the production of rabies vaccines, free of myelin and therefore unable to cause neuro-paralytic accidents, has been choosen in 1950 by Powell and Culbertson [18]. They prepared a vaccine from a fixed strain of rabies virus, adapted to grow in 7-day old embryonated duck eggs. The vaccine is a crude 10% suspension of infected embryos, inactivated by beta-propiolactone and lyophilized. It is actually free from encephalitogenic factors but may cause in some subjects allergic reactions due to avian proteins. This vaccine has been used for many years in the United States. It was, nevertheless, weakly immunogenic and has been abandoned recently. However it is still produced in Switzerland with the introduction of several improvements such as purification and concentration through density gradient centrifugation. The antigenic value of this new purified duck embryo vaccine has been proved to be equal to the value of the best cell culture vaccines [19,20].

2.3 First Attempts to Replicate Rabies Virus in Animal Cell Cultures

The cultivation of rabies virus has been achieved in 1930 by Stoel [21] in primary explants of chick embryo brain; in these cells Stoel could propagate the virus along five serial passages. During the following ten years several other investigators reported the propagation, in cultures of mouse embryo brains, of fixed rabies virus previously passaged in laboratory animals [22]. The first direct in vitro isolation and cultivation of street rabies virus was obtained in 1942 by Plotz and Reagan [23], who from the brain of a human case of rabies and from the brain of a rabid dog, inoculated on primary explants of chick embryo cells, could propagate the virus through eleven and nine transfers, respectively.

Tumor cells have also been used. In 1937, Levaditi and Schoen [24] were able to obtain the replication of street rabies virus in Pearce carcinoma cells grafted into the anterior eye chamber of rabbits. In 1958, Pearson et al.[25] multiplied the fixed rabies virus in the cells of an astrocytoma tumor grafted in C3H mouse.

In 1959 Atanasiu and Lépine[26] propagated street rabies virus, isolated from a fatal case of human rabies after two passages in rabbit brain, in a cell line of mouse glial ependymoma, for up to twenty two passages and observed, in the cytoplasm of the infected cells, the presence of inclusions closely similar to the Negri bodies which are characteristic of street rabies virus in the neuronal cells in the brain of rabid animals. Cultivation of rabies virus in non nervous cell cultures started at the same time after Vieuchange[27,28] described the susceptibility of mouse kidney cell cultures to rabies virus. In 1958, Kissling[29] reported the successful cultivation of fixed rabies virus (obtained from rabies infected mouse brain) and street rabies virus (isolated from salivary glands of a rabid dog) in primary cultures of hamster kidney cells. He could serially propagate the fixed virus through fifteen cell culture passages and the street virus during four passages. Soon after, in 1963, the first experimental cell culture rabies vaccine was obtained by Kissling and Reese[31] in primary hamster kidney cells inoculated with the previously adapted CVS fixed rabies virus. Fenje[30] in 1960 achieved the first adaptation of a rabies virus strain to cell cultures for possible use in vaccine production. He used the strain of rabies virus called SAD, originally isolated from the brain of a rabid dog and serially propagated in mouse brain, and adapted it by means of alternate passages in primary hamster kidney cell cultures and mouse brain. After this adaptation was obtained he could serially propagate the virus in the cells cultures up to the point of obtaining high enough infectivity titers of virus to be used for a vaccine production. Abelseth[32,33] reported in 1964 the propagation of the SAD strain of rabies virus in primary pig kidney cells and its use for the production of an attenuated rabies vaccine for domestic animals. The susceptibility of primary chick embryo cell cultures to the egg-embryo adapted Flury-HEP strain of rabies virus has been reported in 1965 by Kondo[34] who soon after developed with this method an inactivated rabies vaccine for human use.

Cell lines have also been largely used for the cultivation of rabies virus. The BHK/21 cell line, derived from baby hamster kidney by MacPherson and Stoker[35] has been used for the production of a purified concentrated rabies vaccine by Wiktor in 1969[36]; another cell line of hamster origin, the NIL 2 line of Diamond[37] has been selected by Precausta[38] in 1974 for the production at the industrial scale of an inactivated rabies vaccine for veterinary use.

For the production of rabies vaccines for human use, the human diploid cell strain HDCS "Wi 38" developed in 1961 by Hayflick and Moorhead[39] has been selected and used by Wiktor et al. in 1964[40]. This vaccine is now produced in another human diploid cell strain, the MRC-5 developed by Jacob et al.[41].

3 Animal Cell Cultures for Virus Vaccines Production

The in vitro survival and growth of animal cells may be realized in three different ways. In organ cultures an organ, or part of it, often at the embryonic stage, is maintained in a nutrient medium in which its structure and functions are preserved. In tissue cultures, explants are grown in the nutrient medium in vitro, maintaining their cell differentiation, organization and possible functions. In cell cultures the situation is different: the cells from an organ or tissue have been dispersed and made to an homogeneous suspension by means of the action of some proteolytic enzyme (trypsin, pepsin, collagenase, pronase) which disrupts the intercellular bridges. The dispersed cells may form monolayers on the surface of the culture vessels or grow in suspension. Selected cell lines may be derived from a single isolated cell by the cloning procedure. Various types of cell cultures may be used for viral vaccine productions, among them rabies vaccines.

Animal cell cultures are maintained in artificial culture media, of defined formula, made of a balanced salt solution, containing the required amounts of amino-acids and vitamins and supplemented with calf serum, at least during the growing phase of the culture. Antibiotics and antimycotics may be added.

Basically, two types of animal cell cultures may be obtained: primary cultures or cell lines: these may be either of a limited duration of life (diploid cell lines) or illimited (heteroploid continuous cell lines).

3.1 Primary Cell Cultures

Also called cultures of primary explantation these cell cultures are obtained from a cell suspension of an organ or tissue such as chick embryo, baby hamster kidney or fetal bovine embryonic kidney. The cell suspension is distributed in glass or plastic individual containers, on to the surface of which the cells sediment and spread; these primary cells must firmly adhere to the support in order to grow and multiply; they are said to be "anchorage dependent" and cannot multiply if floating in suspension in the culture medium. The adhering cells multiply until they form a complete monolayer regularly covering all the available surface of the container. At that time they stop multiplying because of contact inhibition. During the course of the cell multiplication the culture medium must be changed at variable intervals, depending of the cell density and growth. After that the rich growth medium is replaced by a maintenance medium, without serum; this is usually done when the cell cultures are inoculated with the virus for vaccine production.

Several types of containers may be used for these cell cultures: simple stationary flat bottles, or multiple layers containers and roller bottles designed to increase the surface available to the cells without increasing the volume of the nutrient medium.

3.2 Diploid Cell Lines of Restricted Lifetime

The diploid cell lines are derived from primary cultures of embryonic tissues, for instance human embryo skin or lung tissues, through serial in vitro passages. During these passages the cells, even if originally epithelial, become of the fibroblast type. These cells keep their anchorage dependence and contact inhibition characters. Their kariotype is unaltered: diploid. They do not contain any oncogen and are not transformed. These human diploid cell (H.D.C.) lines are the best suitable for human virus vaccines production, such as the Wi38 HDC line [39] used for rabies vaccine production for more than twenty years, now replaced by the MRC-5 line [41]. But, if these diploid cells may be tranferred during several passages, nevertheless their duration of life is limited: after 40 to 60 passages they die. Therefore, sufficient amounts of primary and secondary seed-lots of cells must be preserved in the frozen state in liquid nitrogen (—180 °C) for sub-cultures.

3.3 Continuous Heteroploid Cell Lines

These cell lines have an unlimited duration of life. They can be serially multiplied through an unlimited number of passages; they represent therefore an unexhaustible source of cell cultures for industrial vaccine production. They most often derive from the spontaneous transformation of a diploid cell line in which, after a given number of passages, foci of transformed cells appear having lost their contact inhibition, forming multilayer sheets of cells, having lost their anchorage dependence and being able to grow in suspension. These transformed cells do not have any longer a normal chromosomic formula: they are called heteroploid. Some of the heteroploid cell lines contain oncogens or have integrated genomic fragments of extraneous viruses. Some of them are oncogenic or tumorigenic when experimentally inoculated into hamsters or nude mice.

Having such an unlimited capacity of reproduction the heteroploid cell lines are the ideal cell substrate for inductrial scale viral vaccines production. Indeed they are used, in practice, for the production of vaccines for veterinary use, such as the continuous baby hamster kidney BHK/21 cell line used for foot-and-mouth disease vaccine and for rabies vaccine productions, from cell cultures obtained in large scale suspensions.

But there exist a great concern about the use of such cells for the production of vaccines for human application, due to their potential oncogenic properties. Basically, such transformed cells cannot be employed for the production of human vaccines, and this has been the international recommendation for many years. Nevertheless, during the past few years, numerous and careful investigations have been carried out to throughfully check and evaluate the oncogenic potential danger of the continuous cell lines and it has eventually been possible to demonstrate that some of them, such the Vero cell line, derived from kidneys of African green monkey *Cercopithecus aethiops* [42], were free of oncogenic properties, at least up to a determined level of serial passages, and that there was no risk for human health in using them for vaccine production [43,44]. Of course, very strict requirements must be fullfilled, specially regarding the amount of cell DNA in the final product, and in any case the cells should be employed only in the production of inactivated vaccines. In these conditions, the continuous cell lines became a good candidate for the large scale production of most needed human viral vaccines. In fact a polio-Vero inactivated vaccine has been developed and produced and an inactivated purified Vero rabies vaccine for human use as well.

3.4 Micro-carriers and Bioreactor Technology

The primary cell cultures, the diploid cell strains and some of the continuous cell lines (for instance Vero) require for growing in vitro the attachment to a solid surface; they are anchorage-dependent, as described by Stoker [45]. This character represents a very stringent limiting factor when large scale cell culture production must be achieved for industrial production of vaccines.

A very elegant solution to this problem has been proposed in 1967 by van Wezel [46]: the use of the small beads of the DEAE-gels, which are maintained in suspension in the culture medium and onto which the cells may attach, grow and multiply. These microcarriers have been used for the cultivation in bioreactors of primary cells and cell strains applied to the production of virus vaccines [47,48]. The particles used as microcarriers must have, according to van Wezel [49] the following characteristics: a moderate positive charge (the cells are negatively charged), a density around 1.05/1.15 to be easily maintained in suspension in the medium with a low speed stirring, a diameter between 150 and 200 nm a smooth surface. Such microcarriers may be used at a concentration of 10.000 particles per ml of culture medium, providing a culture surface of about 10 cm^2. Such a rations between the surface available for cell attachment and the volume of required liquid medium is about then times greater than in stationary flat culture flasks and five time greater than in roller bottles.

The microcarrier cell cultures are performed in reactors where all parameters (pH, temperature, etc) are automatically controlled. Such vessels may have a capacity of several ten to several hundred liters. This new biotechnology has been applied with success to the industrial production of several types of virus vaccines: foot-and-mouth disease vaccine in a pig kidney cell line by Meigner [50], killed poliovaccine in primary monkey kidney cells by van Wezel [51]. Microcarriers and bioreactors have also been used with Vero cells for the industrial production of a killed poliovirus vaccine by Montagnon [52].

4 Cell Culture Rabies Vaccines

Several different types of cell culture rabies vaccines have been developed and tested and are presently available either for human use or for veterinary use.

4.1 Rabies Vaccines for Human Use

Most of the vaccines employed for the prevention of human viral diseases are prepared from attenuated strains of virus: for instance the oral polio vaccine, the vaccines against measles, mumps and rubella. The situation is quite different for rabies vaccines which can be prepated only with completely inactivated virus. No live, modified (attenuated) rabies virus is authorized for human immunization.

4.1.1 Rabies Vaccines Prepared in Primary Cell Cultures of Mammalian Origin

Primary Hamster Kidney Cell Rabies Vaccine

In 1966, Fenje [53] reported the preparation and evaluation in clinical trials in human volunteers of a rabies vaccine prepared in primary hamster kidney cell cultures, with the fixed strain CL-60, concentrated by high speed centrifugation and inactivated with formaldehyde. This vaccine has been licensed for pre-exposure vaccination in 1968 [54].

In the USSR, Selimov and his colleagues have adapted the Canadian SAD strain to hámster kidney cells (Vnukovo-32 strain) and used it for the production of a rabies vaccine since 1971. The vaccine is inactivated by UV irradiation [55,56]. For post-exposure vaccination, 5 ml doses were given daily for 25 d, with boosters after 10, 25 and 30 d [57]. This vaccine is now concentrated and purified by zonal centrifugation in sucrose density gradient and lyophilized and administered in only five doses on days 0, 8, 16, 32 and 60 [58]. It is used in several countries of Eastern Europe.

In China, Lin Fang-Tao and his co-workers have prepared a vaccine from the Beijing strain of fixed rabies virus, adapted to primary hamster kidney cells. The vaccine is inactivated with formalin. It may be adjuvanted with aluminium hydroxyde or concentrated by ultrafiltration and lyophilized [59]. For post-exposure treatment, five doses of 2 ml of concentrated and adjuvanted vaccine are administered on days 0, 7, 14, 24 and 34. This vaccine, produced in the Wuhan Institute for Biologicals, has been proved to be safe and efficient over several years of human use in field conditions [60].

Fetal Bovine Kidney Cell Rabies Vaccine

The fetal bovine kidney cell (FBKC) rabies vaccine has been developed in the Pasteur Institute in Paris by Atanasiu in 1974 [61]. It is prepared with the P. V. (Pasteur Virus) strain of fixed rabies virus, adapted to primary cultures of fetal bovine kidney cells. The vaccine is concentrated and purified by zonal ultra-centrifugation, inactivated by beta-propiolactone and lyophilized [62,63]. This vaccine is now currently used in France and several other countries. For pre-exposure immunization, two doses of 1 ml are given at 4 weeks interval, followed by a booster injection after on year, then every three years [64]. For post-exposure treatment, five doses of 1 ml are administered on days 0, 3, 7, 14 and 30 (a booster injection at day 90 is optional). The tolerance is excellent and the immune response comparable to that obtained with the HDC vaccine given at the same regimen [65,66].

Primary Dog Kidney Cell Rabies Vaccine

The dog kidney cell rabies vaccine has been developed in the Netherland by van Wezel [67] in 1978, using the P.M. (Pitman-Moore) strain of fixed rabies virus. The virus is produced in primary dog kidney cells grown on microcarriers in bioreactors. The vaccine is concentrated by ultrafiltration, inactivated with beta-propiolactone and lyophilizied. It is used for both pre- and post-exposure immunization, according to the regimen adopted for the HDC vaccine, and is well tolerated and quite immunogenic [68,69].

4.1.2 Rabies Vaccines Prepared in Primary Cell Cultures of Avian Origin

- The first vaccine of this type has been developed in Japan by Kondo [70] in 1972. The vaccine is prepared with the Flury HEP (High Egg Passage) strain of rabies virus; it is inactivated with beta-propiolactone and concentrated (ultracentrifugation, ultra-filtration, polyethylene-glycol precipitation). The vaccine is commercially produced in Japan since 1980. For post-exposure treatment 2 ml doses are injected daily for 14 consecutive days.
- In Germany, Barth and his colleagues [71,72] have developed in 1974 a purified chick embryo cell (PCEC) rabies vaccine for human use. The vaccine is prepared with the Flury LEP (Low Egg Passage) strain of rabies virus, concentrated and purified by continuous density gradient centrifugation and inactivated with beta-propiolactone. For post-exposure treatment 6 doses of 1 ml are administered on days 0, 3, 7, 14, 30 and 90; the immune response has been proven to be as good as with the HDC vaccine [73,74,75]. This vaccine appears to be a very promising candidate for a low-cost, highly immunogenic cell culture rabies vaccine for developing countries such as Thailand [76,77] and India [78].
- Another avian embryo cell culture vaccine has been developed by Bektimirova [79] in the USSR, using a rabies virus strain derived from the Pasteur strain, adapted to grow in primary japanese quail embryo cell cultures. The vaccine is concentrated by ultra-centrifugation, inactivated by beta-propiolactone and lyophilized. Two vaccinations regimens have been tested in human volunteers: 2 ml daily for 14 d or 6 injections on days 0, 3, 7, 14, 30 and 90; the second group exhibited higher and longer antibody response.

4.1.3 Rabies Vaccines Prepared in Human Diploid Cell Strains

The cultivation of rabies virus in a human diploid cell strain (Wi-38) achieved by Wiktor in 1964 [40] has led to the preparation of a highly immunogenic rabies vaccine for human use. After being tested in human volunteers by Bahmanyar [80] in 1974 in Iran, where classical nervous tissue vaccines had proved to be unsatisfactory for the post-exposure treatment of persons seriously bitten by rabid wolves, the efficacy of the new human diploid cell (HDC) vaccine was confirmed, in Iran in 1975: 45 persons bitten by two rabid wolves and six rabid dogs were successfully protected by only six 1 ml injections of HDC vaccine given on day 0 (with rabies immune serum), and days 3, 7, 14, 30 and 90 [81].

The high immunogenic activity of the HDC vaccine has been later confirmed in numerous pre- and post-exposure vaccination studies [82]. This vaccine is now produced in MRC-5 cells by the Merieux Institute in France. It is concentrated by ultra-filtration, inactivated with beta-propiolactone and lyophilized. It is widely used in several countries in Europe, Asia and Africa [83] and in the U.S.A. [84].

A similar vaccine is produced in Canada by Connaught Laboratories, in MRC-5 diploid cells, with the CL-77 canadian strain of rabies virus [85].

A non human primate diploid cell line, derived from the lung of a fetal rhesus monkey, has recently been used for the preparation of a rabies vaccine at the Michigan Department of Public Health. This vaccine is inactivated with beta-propiolactone and adjuvanted with aluminium hydroxyde [85,86].

4.1.4 Rabies Vaccines Prepared in Continuous Cell Lines

Human diploid cell and primary kidney cell rabies vaccines are most efficient but their high cost of production remains an obstacle to their use, on a large scale, in the developing countries where they are needed in important quantities. Efforts have therefore been made in order to develop low price, industrial scale, production methods. In this respect, the possibility to produce rabies vaccines in non oncogenic continuous cell lines has constituted a significant progress. The Vero cell line, cultivated on microcarriers in large bioreactors of several hundred liters capacity, are now used for mass production of cell culture rabies vaccines at the Merieux Institute in France [88,89]. This new vaccine has proved to be highly immunogenic and well tolerated in its first clinical trials [90].

4.2 Rabies Vaccines for Veterinary Use

Human rabies vaccines are mainly applied to the post-exposure treatment of persons contaminated through the bite of a rabid animal; their use for preventive immunization is restricted to certain categories of subjects professionnally exposed to the risk of contamination. On the contrary, rabies vaccines for veterinary use are only applied as a measure of medical prophylaxis for the preventive immunization of the animals, either on an individual basis or in large animal populations, sometimes within the frame of mass-vaccination campaings. The amount of vaccine needed for animal vaccinations may therefore be very important and require vaccine production at an industrial scale. For this reason the animal cell culture technology has been applied to the production of rabies vaccines for veterinary use even before to the production of human rabies vaccines. On the other hand it is easier to experiment a newly developed vaccine in animals than to assess the efficacy of a new vaccine in human beings. In particular, the protection conferred by a rabies vaccine in various animal species may be checked by means of the inoculation of a challenge dose of infective rabies virus; this is obviously impossible in man.

Another difference must be considered: for human vaccination only inactivated vaccines may be used; on the contrary, for the vaccination of animals both live attenuated and inactivated rabies vaccines may be employed.

4.2.1 Modified Live Virus Rabies Vaccines for Veterinary Use

The egg-embryo adapted Flury-HEP strain is used, in the USA, for the production in dog kidney cells of an atenuated vaccine for dogs and cats.

Most other attenuated vaccines are prepared with virus strains derived from the SAD strain of Abelseth [32,33], on dog kidney cells or bovine kidney cells. One of the SAD derived strains: the ERA strain, is used in Canada for the production in primary pig kidney cells of a potent vaccine for dogs and domestic herbivores [91 92]. Another strain obtained from the SAD in USSR: the Vnukovo-32 strain, is used for the production in hamster kidney cells of a live modified vaccine for domestic carnivores and herbivores [93].

Recently, modified live rabies virus vaccines have been experimented for the oral immunization of wild animals. When other measures, such as the reduction of fox populations density had proved to be relatively inefficient in controlling the enzootic of fox rabies, several laboratory and field studies on the oral vaccination of foxes have provided very encouraging results. The first studies were done in Switzerland by Steck and his colleagues [94 95] using the SAD virus strain. Further field trials were then realized in Germany by Wachendorfer [96] and Schneider [97], with a vaccine produced in BHK cells with the SAD-B19 clone of the virus.

4.2.2 Inactivated Rabies Vaccines for Veterinary Use

One of the advantages of the live modified rabies vaccines is that they are able to induce a long lasting immunity after a single injection. But, in some sensitive animal species, like cats, they may be responsible for cases of vaccine induced rabies. Therefore inactivated vaccines which are absolutely free of such a risk must be preferred. Indeed they are the only authorized vaccines in many countries. On the other hand, potent inactivated cell culture rabies vaccines are also able to immunize for long periods of time, specially when associated with adjuvants of immunity [98].

These vaccines are produced in hamster kidney cell lines, such as the NIL-2. The virus is inactivated with beta-propiolactone [99]. The product may be freeze-dried or in the liquid form if associated to adjuvants of immunity such as aluminium hydroxyde or saponin [100].

Rabies vaccines for domestic animals are often associated to vaccines against other animal diseases: foot-and-mouth disease vaccine in cattle; canine distemper, hepatitis and leptospirosis vaccines in dogs; panleucopenia vaccine in cats.

5 New Generation of Rabies Vaccines

The cell culture biotechnology has made possible the production of rabies vaccines constituted of inactivated purified rabies virions.

These bullet shaped enveloped virus particles, 180 nm by 75 nm, contain a single non segmented negative strand RNA and five structural proteins: the nucleoprotein (N), the nucleo-capsid associated protein (NS), the virion transcriptase (L), the matrix (M) protein of the lipoprotein envelope, and the glycoprotein (G) which forms the spikes protruding on the external surface of the virus membrane.

Among the five viral structural proteins, the glycoprotein (G) is the only one which is able to induce the production of virus neutralizing antibodies and the resistance to rabies virus infection [101]. It has been therefore considered logical to attempt immunization against rabies with a sub-unit vaccine made of purified glycoprotein [102]; such a sub-unit vaccine being free of the virus genome. Unfortunately the immunizing activity of the solubilized glycoprotein is weak [103]; but it can be increased by the incorporation of the glycoprotein into lipid vesicles (virosomes) [104] or, better, by the anchorage of the glycoprotein molecules onto the surface of preformed lipid vesicles called immunosomes [105,106].

Further progress in the knowledge of the amino acid sequence of the glycoprotein has made possible to determine which peptide fragments were responsible for the induction of the rabies neutralizing antibodies [107], as well as of the cellular immunity and synthetic peptides have been produced [108] which could be the first step for the production of synthetic rabies vaccines.

Another approach has concerned the possibe use of anti-idiotypic antibodies as an antigen-independent immunizing agent: murine monoclonal antibodies, specific of the immunogenic antigenic determinants of the glycoprotein, induce when injected into rabbits the production of anti-idiotypic antibodies which mimic the original antigenic structure and induce, when injected into mice, the production of a specific virus neutralizing antibody response [109,110].

Presently the synthetic peptides and the antiidiotypes are not yet ready for practical use at least in the near future. The genetic engineering offers much more promising possibilities. A double stranded cDNA copy of the messenger mRNA of the virus coding for the rabies glycoprotein has been isolated and cloned and its complete nucleotid sequence determined. The clone has been inserted into a plasmid and by transfection a recombinant vaccinia virus has been obtained which had incorporated the rabies glycoprotein-coding sequence into its own genome [111]. Cultures of BHK-21 cells, inoculated with the vaccinia recombinant virus have expressed rabies glycoprotein both on their surface and in their cytoplasm. Mice inoculated with the vaccinia virus recombinant, by scarification or by footpad inoculation have developed high titers of neutralizing antibodies and resisted a subsequent intra-cerebral challenge with street rabies virus [112]. The same induction of neutralizing antibodies and of resistance to challenge has been obtained with inactivated preparations of BHK-21 infected cell extracts, purified vaccinia virus recombinant, as well as with purified glycoprotein isolated from the infected cells. Furthermore, mice given the vaccinia virus recombinant as a vaccine by the oral route developed resistance to subsequent challenge with rabies virus [113]; this could be applied to the immunization of wildlife by the oral route [114].

6 Conclusion

Rabies remains today, in many parts of the World, an important zoonose which endangers both human and animal health. The control of rabies may be achieved, besides the implementation of strict measures of sanitary prophylaxis, by the large use of medical prophylaxis i.e. immunization. The vaccination against rabies may be applied at three different levels: to the animal species which constitute the reservoir of the disease (wildlife and stray dogs); to the domestic animals, mainly dogs and cats, which constitute the transmission link between the reservoir animals and the human population; to the human beings exposed to the contamination from a rabid animal.

The production of rabies vaccines has come into the modern stage with the introduction of the animal cell cultures technology. This technology has brought dramatic improvements in the quality of both human and animal rabies vaccines. The new cell culture rabies vaccine are now perfectly safe, well tolerated, highly immunogenic and satisfy the most stringent international requirements. They may be produced at an industrial scale in large bioreactors, using continuous cell lines and microcarrier cell cultures. The possible use of a vaccinia virus recombinant containing the rabies virus glycoprotein gene might bring exciting new achievements in the production of rabies vaccines and help man to eventually win his fight against rabies.

7 References

1. Steele, J. H.: History of rabies, in: The Natural History of Rabies (Baer, G. M., ed.) p. 1, New York—San Francisco—London, Acad. Press 1975
2. World Survey of Rabies XXI, WHO/Rabies/84.195 (ed.) World Health Organization, Geneva, 1984
3. Galtier, P. V.: C. R. Acad. Sci. (Paris) *93*, 284 (1881)
4. Pasteur, L.: ibid. *92*, 1259 (1881)
5. Pasteur, L.: ibid. *98*, 457 (1884)
6. Pasteur, L.: ibid. *98*, 1229 (1884)
7. Pasteur, L.: ibid. *101*, 765 (1885)
8. Högyes, A.: Orv. Hetil. *31*, 121 (1887)
9. Puscariu, E., Vesesco, E.: Ann. Inst. Pasteur *9*, 210 (1895)
10. Babès, V.: Traité de la Rage, Paris, Baillères et Fils (1912)
11. Fermi, C.: Z. Hyg. Infektionskr. *58*, 233 (1908)
12. Semple, D.: Sci. Mem. Med. Sanit. Dep. India, n° *44* (1911)
13. Hempt, A.: Ann. Inst. Pasteur *39*, 362 (1925)
14. Lépine, P., Atanasiu, P., Gamet, A., Vialat, C.: Phenolized, freeze dried sheep brain vaccine. Method used at the Pasteur Institute, Paris, in: Laboratory Techniques in Rabies (Kaplan, M. M. & Koprowski, H. eds.) p. 204, Geneva, W.H.O., 1973[3]
15. Fuenzalida, E., Palacios, R.: Biol. Inst. Bacteriol., Chile *8*, 3 (1955)
16. Svet-Moldavskij, G. J. A., Andjaparidze, O. G., Unanov, S. S., Karakajumcan, M. K., Svet-Moldavskaja, I. A., Mucnik, L. S., Hieninson, M. A., Ravkina, L. I., Mtvarelidze, A. A., Volkova, O. F., Kriegshaber, M. R., Kalinkina, A. G., Salita, T. V., Klimovickaja, V. I., Bondaletova, I. N., Rojhel, V. M., Kiseleva, I. S., Levcenko, E. N., Marennikova, S. S., Leonidova, S. L.: An Allergen-Free Antirabies Vaccine. Bull. World Health Organization *32*, 47 (1965)
17. Gispen, R., Saathof, B.: Arch. gesamte Virusforsch. *15*, 377 (1965)
18. Powell, H. M., Culbertson, C. G.: Pub. Health Rep. *65*, 400 (1950)
19. Glück, R., Keller, H., Mischler, R., Wegmann, A., Germanier, R.: New aspects concerning the immunogenicity of rabies vaccines produced in animal brains (Duck-embryo), in: Rabies in the Tropics (Kuwert, E., Mérieux, C., Koprowski, H., Bögel, K. eds.) p. 181, Berlin—Heidelberg—New York—Tokyo, Springer-Verlag 1985
20. Wegmann, A., Glück, R., Keller, H., Matthieu, J. M.: Schweiz. Medizin. Wochensch. *49*, 115 (1985)
21. Stoel, G.: C. R. Soc. Biol. *104*, 851 (1930)
22. Wiktor, T. J., Clarck, H. F.: Growth of rabies virus in cell cultures, in: The Natural History of Rabies (Baer, G. M. ed.) p. 115, New York—San Francisco—London, Acad. Press 1975
23. Plotz, J., Reagan, R.: Science *95*, 102 (1942)
24. Levaditi, C., Schoen, R., Rénié, L.: Ann. Inst. Pasteur *58*, 353 (1937)
25. Pearson, H. E., Atanasiu, P., Lépine, P.: ibid. *94*, 1 (1958)
26. Atanasiu, P., Lépine, P.: ibid. *96*, 72 (1959)
27. Vieuchange, J., Vialat, C., Gruest, J., Bequignon, R.: ibid. *90*, 361 (1956)
28. Vieuchange, J., Bequignon, R., Gruest, J., Vialat, C.: Bull. Acad. Natl. Méd. Paris, 5/6, *77* (1958)
29. Kissling, R. E.: Proc. Soc. Exp. Biol. Med. *98*, 223 (1958)
30. Fenje, P.: Canad. J. Microbiol. *6*, 479 (1960)

31. Kissling, R. E., Reese, D. R.: J. Immunol. *91*, 362 (1963)
32. Abelseth, M. K.: Canad. Vet. J. *5*, 84 (1964)
33. Abelseth, M. K.: ibid. *5*, 279 (1964)
34. Kondo, A.: Virology *27*, 199 (1965)
35. MacPherson, I., Stoker, M.: ibid. *16*, 147 (1962)
36. Wiktor, T. J., Sokol, F., Kuwert, E., Koprowski, H.: Proc. Soc. Exp. Biol. Med. *131*, 799 (1969)
37. Diamond, L.: Int. J. Cancer *2*, 143 (1967)
38. Precausta, P.: Symp. Series immunobiol. Standard *21*, 162 (1974)
39. Hayflick, L., Moorhead, P. S.: Exp. Cell Res. *25*, 585 (1961)
40. Wiktor, T. J., Fernandes, M. V., Koprowski, H.: J. Immunol. *93*, 353 (1964)
41. Jacob, J. P., Jones, C. M., Baille, J. P.: Nature *277*, 168 (1970)
42. Rhim, J. S., Schell, K., Creasy, B., Case, W.: Proc. Soc. Exp. Biol. Med. *132*, 670 (1969)
43. Levenbook, I. S., Petricciani, J. C., Elisberg, B. L.: J. Biol. Standard *12*, 391 (1984)
44. Levenbook, I. S., Petricciani, J. C., Qi, Y., Elisberg, B. L., Rogers, J. L., Jackson, L. B., Wierenga, D. E., Webster, B. A.: ibid. *13*, 135 (1985)
45. Stoker, M., O'Neill, C., Berryman, S., Waxman, V.: Int. J. Cancer *3*, 683 (1968)
46. Van Wezel, A. L.: Nature *216*, 64 (1967)
47. Van Wezel, A. L.: Progr. immunobiol. Standard *5*, 187 (1972)
48. Van Wezel, A. L.: Microcarrier cultures of animal cells, in: Tissue Culture methods and application (Kruse, P. F., Patterson, M. K. eds.) p. 372, New York—London, Acad. Press 1973
49. Van Wezel, A. L.: Develop. biol. Standard *37*, 143 (1976)
50. Meigner, B.: ibid. *42*, 141 (1979)
51. Van Wezel, A. L., van der Velden-de Groot, C. A. M., van Herwaarden, J. A. M.: ibid. *46*, 151 (1980)
52. Montagnon, B. J., Fanget, B., Nicolas, A. J.: ibid. *47*, 55 (1981)
53. Fenje, P., Pinteric, L.: Amer. J. Pub. Hlth *56*, 2106 (1966)
54. Fenje, P.: Symp. Series immunobiol. Standard *21*, 148 (1974)
55. Selimov, M. A., Aksenova, T. A.: ibid. *1*, 377 (1966)
56. Selimov, M. A., Aksenova, T., Gribencha, L., Kljueva, E., Guliev, M., Mirozoeva, S., Kocofane, V., Anina-Radcenko, N., Markkarjan, A., Stepanova, I., Kuznecova, A., Presnecova, N., Andreeva, S., Fursova, A., Soluha, E.: ibid. *21*, 179 (1974)
57. Selimov, M. A., Aksenova, T., Klyueva, E., Gribencha, L., Lebedeva, I.: ibid. *40*, 57 (1978)
58. Selimov, M. A., Klyueva, E. V., Aksenova, T. A., Lebedeva, I. R., Gribencha, L. F.: ibid. *40*, 141 (1978)
59. Lin Fang-Tao, Zeng Fan-zhen, Lu Long-mu, Lu Xiao-Zen, Zen Rongfang, Yu Yongrin, Chang Nai-min: J. Infect. Dis. *147*, 467 (1983)
60. Lin Fang-Tao, Lu Xiao-zeng, Cheng Shu-beng, Wang Guao-fu, Zeng Fan-zhen, Lu Long-mu, Cheng Nai-min, Fang Ji-zui: Further study on the stability and efficacy of the primary hamster kidney cell rabies vaccine, in: Improvements in rabies post-exposure treatment (Vodopija, I., Nicholson, K. G., Smerdel, S., Bijok, U. eds.) p. 37, Zagreb Institute of Public Health 1985
61. Atanasiu, P., Tsiang, H., Gamet, A.: Ann. Microbiol. (Inst. Pasteur) *125B*, 419 (1974)
62. Atanasiu, P., Tsiang, H., Lavergne, M., Chermann, J. C.: ibid. *128B*, 297 (1977)
63. Atanasiu, P., Tsiang, H., Reculard, P., Aguilon, F., Lavergne, M., Adamowicz, Ph.: Develop. biol. Standard. *40*, 35 (1978)
64. Guesry, P., Sureau, P., Lery, L., Cerisier, Y., Gamet, A., Atanasiu, P.: Comparative clinical trial of three different rabies vaccines used in pre-exposure vaccination, in: Cell Culture Rabies Vaccines and their protective Effect in Man (Kuwert, E. K., Wiktor, T. J., Koprowski, H. eds.) p. 314, Geneva, International Green Cross 1981
65. Sureau, P., Rollin, P. E., Loucq, C.: Ann. Virol. (Inst. Pasteur) *135E*, 277 (1984)
66. Sureau, P., Rollin, P. E., Fritzell, B., Loucq, C., Marie, F. N., Courrier, A., Simonnet, Ph., Arbogast, J., Fremont, J., Gaudiot, C., Malo, J. P.: Reactogenicity and immunogenicity of primary fetal bovine kidney cells (FBKC) rabies vaccine: post-exposure treatment in: Improvements in rabies post-exposure treatment (Vodopija, I., Nicholson, K. G., Smerdel, S., Bijok, U. eds.) p. 47, Zagreb Institute of Public Health 1985

67. Van Wezel, A. L., Van Steenis, G.: Develop. biol. Standard. *40*, 69 (1978)
68. Van Steenis, G., Van Wezel, A. L., Van der Marel, P. J., Hannik, Ch. A.: Dog kidney cell rabies vaccine: some aspects of its control and efficacy in man, in: Cell culture rabies vaccines and their protective effect in man (Kuwert, E., Wiktor, T. J., Koprowski, H. eds.) p. 78, Geneva, International Green Cross 1981
69. Van Steenis, G., Van Wezel, A. L., Hannik, Ch. A., Van der Marel, P., Osterhaus, A. D. M. E., de Grott, I. G. M., Koning, C., Lelyveld, J. L. M., Ruitenberg, E. J.: Immunogenicity of dog kidney cell rabies vaccine (DKCV), in: Rabies in the Tropics (Kuwert, E., Mérieux, C., Koprowski, H., Bögel, K. eds.) p. 172, Berlin—Heidelberg—New York—Tokyo, Springer-Verlag 1985
70. Kondo, A., Takashima, Y., Suzuki, M.: Symp. Series immunobiol. Standard *21*, 182 (1974)
71. Barth, R., Gruschkau, H., Bijok, U., Hilfenhaus, J., Hinz, J., Milcke, L., Moser, H., Jaeger, O., Ronneberger, H., Weinmann, E.: J. Biol. Stand. *12*, 29 (1984)
72. Barth, R., Gruschkau, H., Jaeger, O., Milcke, L., Weinmann, E.: Behring Inst. Mitt. *76*, 142 (1984)
73. Bijok, U., Barth, R., Gruschkau, H., Vodopija, I., Smerdel, S., Kukla, H.: Clinical trials in healthy volunteers with the new purified chick embryo cell rabies vaccine in man, in: Rabies in the Tropics (Kuwert, E., Mérieux, C., Koprowski, H., Bögel, K. eds.) p. 125, Berlin—Heidelberg—New York—Tokyo, Springer-Verlag 1985
74. Bijok, U.: Purified chick embryo cell (PCEC) rabies vaccine: a review of clinical development 1982–1984, in: Improvements in rabies post-exposure treatment (Vodopija, I., Nicholson, K. G., Smerdel, S., Bijok, U. eds.) p. 103, Zagreb Institute of Public Health 1985
75. Vodopija, I., Smerdel, S., Bijok, U.: P.C.E.C. rabies vaccine and HRIG in post-exposure protection against rabies, in: Rabies in the Tropics (Kuwert, E., Mérieux, C., Koprowski, H., Bögel, K. eds.) p. 133, Berlin—Heidelberg—New York—Tokyo, Springer-Verlag 1985
76. Wasi, C., Chaiprasithikul, P., Puthavathana, P., Chavanich, L., Thongcharoen, P.: Immunogenicity and reactogenicity of the new tissue culture rabies vaccine for human use (Purified chick embryo cell culture), in Improvements in rabies post-exposure treatment (Vodopija, I., Nicholson, K. G., Smerdel, S., Bijok, U. eds.) p. 85, Zagreb Institute of Public Health 1985
77. Wasi, C., Chaiprasithikul, P., Chavanich, L., Puthavathna, P., Thongcharoen, P., Trishananandа, M.: Lancet *40*, 8471 (1986)
78. Sehgal, S.: Report on the trials of the P.C.E.C. (Purified chick embryo cell) rabies vaccine in India, in: Improvements in rabies post-exposure treatment (Vodopija, I., Nicholson, K. G., Smerdel, S., Bijok, U. eds.) p. 71, Zagreb Institute of Public Health 1985
79. Bektimirova, M. S., Pille, E. R., Matevosyan, K. SH., Nagieva, F. G.: Acta Virol. *27*, 59 (1983)
80. Bahmanyar, M.: Symp. Series immunobiol. Standard. *21*, 231 (1974)
81. Bahmanyar, M., Fayaz, A., Nour-Salehi, S., Mahammadi, M., Koprowski, H.: J. Amer. Med. Ass. *236*, 2571 (1976)
82. Wiktor, T. J., Plotkin, S. A., Koprowski, H.: Develop. biol. Standard. *40*, 3 (1978)
83. Roumiantzeff, M., Montagnon, B., Vincent-Falquet, J. C., Bussy, L., Charbonnier, C.: Rapport sur l'utilisation du vaccin rabique préparé sur culture de cellules diploïdes humaines pour l'immunisation avant et après exposition, in: Rabies in the Tropics (Kuwert, E., Mérieux, C., Koprowski, H., Bögel, K. eds.) p. 91, Berlin—Heidelberg—New York—Tokyo, Springer-Verlag 1985
84. Winkler, W. G.: Current status of use of human diploid cell strain rabies vaccine in the U.S., in: Improvements in rabies post-exposure treatment (Vodopija, I., Nicholson, K. G., Smerdel, S., Bijok, U. eds.) p. 3, Zagreb Institute of Public Health 1985
85. Johnson, S. E., Pearson, E. W., Ing, W. K.: Clinical responses in humans to rabies vaccine prepared in MRC-5 diploid cells from canadian seed virus, in: Rabies in the Tropics (Kuwert, E., Mérieux, C., Koprowski, H., Bögel, K. eds.) p. 99, Berlin—Heidelberg—New York — Tokyo, Springer-Verlag 1985
86. Berlin, B. S., Mitchell, J. R., Burgoyne, G. H., Oleson, D., Brown, W. E., Goswick, C., McCullough, N. B.: J. Amer. Ass. *247*, 1776 (1982)
87. Burgoyne, G. H., Kajiya, K. D., Brown, D. W., Mitchell, J. R.: J. Infect. Dis. *152*, 204 (1985)
88. Montagnon, B. J., Fournier, P., Vincent-Falquet, J. C.: Un nouveau vaccin antirabique à usage

humain, in: Rabies in the Topics (Kuwert, E., Merieux, C., Koprowski, H., Bögel, K. eds.) p. 138, Berlin—Heidelberg—New York—Tokyo, Springer-Verlag 1985
89. Fournier, P., Montagnon, B., Vincent-Falquet, J. C., Ajjan, N., Drucker, J., Roumiantzeff, M.: A new vaccine produced from rabies virus cultivated on Vero cells, in: Improvements in rabies post-exposure treatment (Vodopija, I., Nicholson, K. G., Smerdel, S., Bijok, U. eds.) p. 115, Zagreb Institute of Public Health 1985
90. Svjetlicic, M., Vodopija, I., Smerdel, S., Ljibicic, M., Baklaic, Z., Vincent-Falquet, J. C., Pouradier-Duteil, X., Zayet-Bechelet, M.: Compatibility and immunogenicity of Merieux's purified vero rabies vaccine (PVRV) in healthy subjects, in: Improvements in rabies post-exposure treatment (Vodopija, I., Nicholson, K. G., Smerdel, S., Bijok, U. eds.) p. 123, Zagreb Institute of Public Health 1985
91. Abelseth, M. K.: Symp. Series immunobiol. Standard. *1*, 367 (1966)
92. Abelseth, M. K.: Canad. Vet. J. *8*, 221 (1967)
93. Selimov, M. A.: World Health Organization, V.P.H./RAB.RES./73 I. Add. 1 (1973)
94. Steck, F., Wandeler, A., Bichsel, P., Capt, S., Haflinger, U., Schneider, L.: Comp. Immun. Microbiol. Infect. Dis. *5*, 165 (1982)
95. Steck, F., Wandeler, A., Bichsel, P., Capt, S., Schneider, L. G.: Zbl. Vet. Med. B *29*, 372 (1982)
96. Wachendorfer, G., Frost, J., Gutman, B., Eskens, U., Schneider, L. G., Dingeldein, W., Hofmann, J.: Rev. Ecol. (Terre et Vie) *40*, 257 (1985)
97. Schneider, L. S., Cox, J. H.: Tierärztl. Umsch. *38*, 315 (1983)
98. Blancou, J.: Rev. sci. tech. Off. int. Epiz. *4*, 235 (1985)
99. Petermann, H. G., Lang, R., Branche, R., Soulebot, J. P., Mackowiak, C.: C. R. Acad. Sci. Paris, *265*, 2143 (1967)
100. Petermann, H. G., Soulebot, J. P., Lang, R., Branche, R.: ibid. *270*, 234 (1970) (1970)
101. Wiktor, T. J., György, E., Schlumberger, H. D., Sokol, F., Koprowski, H.: J. Immunology *110*, 269 (1973)
102. Crick, J., Brown, F.: Nature (London) *222*, 92 (1969)
103. Brown, F., Crick, J.: Symp. Series Immunobiol. Standard *21*, 119 (1974)
104. Cox, J. H., Dietzschold, B., Weiland, F., Schneider, L. G.: Infection and Immunity *30*, 572 (1980)
105. Perrin, P., Thibodeau, L., Dauguet, C., Fritsch, A., Sureau, P.: Ann. Virol. (Inst. Pasteur) *135E*, 183 (1984)
106. Perrin, P., Thibodeau, L., Sureau, P.: Vaccine *3*, 325 (1985)
107. Dietzschold, B., Wiktor, T. J., MacFarlan, R., Varrichio, A.: J. Virology *44*, 595 (1982)
108. MacFarlan, R., Dietzschold, B., Wiktor, T. J., Kiel, M., Houghten, R., Lerner, R. A., Sutclife, J. G., Koprowski, H.: J. of Immunology *133*, 2748 (1984)
109. Reagan, K., Wunner, W. H., Wiktor, T. J., Koprowski, H.: J. of Virology *48*, 660 (1983)
110. Reagan, K. J.: Acta Virologica *29*, 15 (1985)
111. Kieny, M. P., Lathe, R., Drillien, R., Spehnert, S., Skory, S., Schmitt, D., Wiktor, T. J., Koprowski, H., Lecocq, J. P.: Nature *312*, 163 (1984)
112. Lathe, R., Kieny, M. P., Lecocq, J. P., Drillien, R., Wiktor, T. J., Koprowski, H.: Immunization against Rabies Using a Vaccinia-Rabies Recombinant Virus Expressing the Surface Glycoprotein. in Vaccines 85 (Lerner, R. A., Chanock, R. M., Brown, F. eds.) p. 157, Cold Spring Harbor Laboratory 1985
113. Wiktor, T. J., MacFarlan, R. I., Reagan, K. J., Dietzschold, B., Curtis, P. J., Wunner, W. H., Kieny, M. P., Lathe, R., Lecocq, J. P., Mackett, M., Moss, B., Koprowski, H.: Proc. Natl. Acad. Sci. U.S.A. *81*, 7194 (1984)
114. Blancou, J., Kieny, M. P., Lathe, R., Lecocq, J. P., Pastoret, P. P., Soulebot, J. P., Desmettre, P.: Nature, *322*, 373 (1986)

The Use of BHK Suspension Cells for the Production of Foot and Mouth Disease Vaccines

P. J. Radlett
Coopers Animal Health Limited, Biologicals Production Laboratory, Ash Road, Pirbright, Woking, Surrey, GU24 ONQ, England

The isolation of Baby Hamster Kidney (BHK) 21 cells and their adaptation, characterisation and development for the commercial manufacture of foot and mouth disease vaccines over a twenty year period is reviewed, including the preparation of media, handling of cell substrates and control of the physical environment during manufacture. A description is given of the successful operation of the BHK process which depends on a series of manufacturing stages, all of which make an essential contribution to the final quality of the product and are complementary to the ultimate assessment of quality as defined by final product testing.

Although the process has been operated now for many years the exciting developments in molecular biology and genetic engineering are contributing to our knowledge and understanding of the existing process and may provide new approaches for the future.

1 Introduction

Foot and Mouth Disease (FMD) is recognised as being the most economically significant animal virus disease in the world and its prevention/control and ultimate eradication are recognised as prime objectives for all major animal health programmes. The virus occurs as seven distinct serological types and within each major type many distinct strains occur with greater or lesser significance in different regions and new strains emerge continually. This composite picture gives rise to a current world demand for around 1000 million equivalent monovalent doses of vaccine, of which approximately 50% is manufactured from BHK suspension cells.

Early large scale production of Foot and Mouth Disease vaccines was achieved using primary epithelial bovine tongue tissue [1]. Although still used today in some laboratories the Frenkel process suffers from the limited availability of suitable epithelial tissue and from the risks of introducing adventitious viral and other agents common to all primary tissue culture systems.

Early in the 1960s there was interest in the production of attenuated FMD vaccines, produced for example in baby mice. Within a few years doubts were being expressed internationally as to the safety of such vaccines for FMD and attention was once again focussed on inactivated vaccines. At about the same time progress in tissue culture techniques was beginning to offer the basis for a commercial alternative to the Frenkel process.

The advent of continuous cell lines offered great potential for vaccine manufacture, and the demonstration by workers at The Animal Virus Research Institute (AVRI), Pirbright, England that a number of FMD strains could be grown in baby hamster kidney cells (BHK Strain 21 Clone 13) marked a significant point in the development of Foot and Mouth Disease Vaccines. This work was soon to be followed by the adaptation, or selection, of a subline of these anchorage dependent BHK 21 cells which grew freely in suspension culture, thus providing for the first time the opportunity of a process which was readily amenable to scale-up and not limited by the availability of suitable primary tissue or by the constraints of an anchorage dependent tissue culture system. Cell culture, and hence virus vaccine production could now be undertaken in the hitherto unprecedented volumes required for the effective control of the disease.

2 The Cell

BHK 21 Cl 13 cells are a continuous line of hamster kidney cells derived in 1962 by Macpherson and Stoker [2]. The original isolates were pseudodiploid and anchorage dependent, but nevertheless possessed the property of an apparently indefinite life in culture by serial subcultivation. Perhaps somewhat surprisingly these cells were shown to be susceptible to a range of strains of Foot and Mouth Disease virus [3] and because of these characteristics were subjected to intensive study as a potential substrate for vaccine production. The anchorage dependent growth characteristic placed a severe limitation on the potential for scale-up in the conventional culture systems generally available at the time, giving Capstick et al. [4] the incentive to adapt or select sub-lines from the original population with the ability to grow freely in suspension.

These aneuploid sub-lines have been well characterised [5] and cells derived from stored material produced during these early studies remain the basic substrate used for the production of a large proportion of the Foot and Mouth Disease vaccine manufactured today. These suspension sub-lines have been shown to be non-tumorigenic [5] and in more recent times have been screened in accordance with the requirements of the United States Department of Agriculture (U.S.D.A.) and shown to be free from specific contaminating agents including bacteria, viruses and mycoplasma [6]. It should be emphasised that these comments apply only to sub-lines tested, stored and managed under closely controlled conditions and it cannot be assumed that sub-lines obtained from the various commercial sources now available are free from such agents unless the banks have been specifically tested.

Since these early studies many workers have published findings describing the various metabolic requirements, physical characteristics, and properties of BHK suspension cells and in a number of cases characteristics described in one laboratory cannot be reproduced in another. Over the years there have been many reports of different properties arising in culture and there are now many variant cell populations stored in different laboratories, all derived from the original isolate and all known as "BHK suspension cells".

Whether different characteristics arise as a result of mutation under the different selective pressures exerted in different laboratories, or whether the original parent population was sufficiently heterogenous already to include at low frequency cells with the variety of different characteristics subsequently reported remains an open question. In either case it seems reasonable to conclude that minor qualitative differences in local media components and small differences in cell handling techniques will in many cases favour the emergence of sub-lines with characteristics different from the original parent population.

To illustrate the differences which can occur, the following examples are cited:

1) The media permitting enhanced growth published by Zoletto and Gagliardi [7] required further supplementation before enhanced growth was seen with the cell sub-line grown by Telling and Radlett [8].
2) The BHK suspension cells handled at the laboratory in Brescia may be grown both on glass and in free suspension culture [9]. In contrast the cells propagated by AVRI at Pirbright will survive but not flatten or exhibit the characteristic fibroblastic behaviour of the anchorage dependent lines on glass. The transience of this characteristic was recently demonstrated by transferring cells from Pirbright to Birmingham University, where conditions were such that the cells were found to exhibit typical fibroblastic behaviour [10].
3) The cloned sub-line designated AC 9 by Spier et al. [11] and shown to have enhanced susceptibility to one particular strain of FMD virus lost this characteristic when passaged and stored in our laboratories. On continued serial passage in industrial scale culture systems, this transient loss was reversed and the cells again showed enhanced susceptibility [12]. Although Spier found this cloned sub-line was stable, the above finding is consistent with his general hypothesis that the relative numbers of susceptible and unsusceptible cells in each population determine the overall susceptibility of that population [13].
4) Variations in the susceptibility of BHK suspension cells from a number of different sources were described by Clarke and Spier [13] who used the spectrum of FMD

viruses to which the cells were susceptible to characterise the cell sub-lines. More recently differences have been shown between stocks of cells held by AVRI and by our laboratory at Pirbright and similar differences have been demonstrated between cells held by commercial tissue culture suppliers and both of the above sources. Unfortunately, and contrary to the findings of Spier, in our hands no one cell sub-line appears universally more susceptible or productive than another, but enhanced susceptibility to some strains is countered by reduced susceptibility to others.

The above comments demonstrate some of the difficulties facing a manufacturer required to produce a range of antigens from his cell substrate. Selecting a sub-line for its productivity of a single virus strain, although possible, is not very practical, and a sub-line must be chosen or developed which has high productivity for most of the virus strains required. Having identified such a substrate great care must be taken to retain these desirable characteristics. Thus, it is of critical importance to manage the cell production regime so as to ensure that only materials of uniform and controlled quality are used, and to operate a strict revival policy to maintain the production life of the cell substrate within defined limits. Despite the apparent complexities, the simple operating procedure described below has shown that maintenance of desirable characteristics can be easily achieved in a well organised laboratory.

Table 1. Cell storage system. Coopers Animal Health Ltd. — 1986

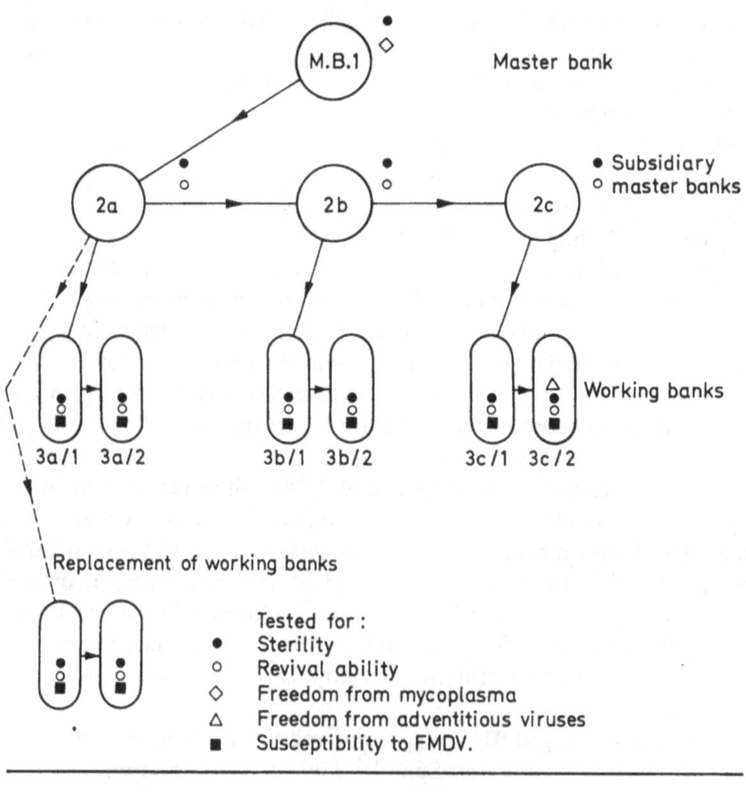

The cell storage and revival procedure used in our own laboratories is shown in Table 1. Working banks are produced as required from tested master banks, and subjected to a further series of quality control tests. Fresh production revivals are made from these working banks at regular intervals, and their performances assessed on a weekly basis against defined quality control procedures. At the end of a 10 week period, or sooner if unsatisfactory quality control results are obtained, fresh revivals are made. This regime minimises the risk of variant cell populations emerging, and eliminates their persistence beyond the 10 week culture period.

3 The Medium

Traditionally, BHK cells have been grown in Eagle's medium, modified to contain twice the original concentration of amino acids and vitamins [2] with the addition of a percentage of animal serum and usually some other ill-defined protein digest such as peptone or tryptose phosphate broth [14,15].

Supplementation studies in depleted media led to the development of media capable of supporting the growth of cells to densities up to 6.5×10^6 per ml [16] by addition of extra glutamine, glucose and vitamins. This medium was further developed in our laboratory (Draper — personal communication) by addition of yeast extract and still more glutamine to produce a medium capable of routinely supporting the growth of over 8×10^6 cells per ml.

In developing improved media it must be remembered that ultimately it is product formation and not cell density that is of prime concern. In this example growth to very high cell densities resulted in the production of cells with significantly reduced susceptibility to virus. As a result of these findings, maximum cell densities in the range $3\text{–}4 \times 10^6$ per ml from the cell growth stage are generally favoured for commercial manufacture of FMD virus [17].

Other workers adopted a different approach and have developed inexpensive, simplified, or serum-free media for the growth of BHK cells [18–20]. Other conventional media such as Roswell Park Memorial Institute (RPMI) have also been shown to give satisfactory results [10]. Although a considerable volume of work has been published on media development, relatively few systematic studies have examined the quantitative nutritional requirements of BHK cells. Arathoon and Telling [21] attempted to obtain such information in their study of the utilisation of glucose and amino acids. They found that the amino acids used in greatest quantity were glutamine and glutamate acid, leucine, asparagine, serine, isoleucine and arginine, but cystine, glutamine and glutamic acid were almost completely depleted by the cells and were considered to be the amino acids most likely to limit cell growth. Only 70% of the amino acids utilised could be accounted for by incorporation to cellular protein and some of the amino acids, particularly glutamine and glutamate may be used as an energy source, even in the presence of glucose. This finding has been confirmed by Butler using microcarrier cultures of anchorage dependent BHK cells [22]. The same author extended the study [23] to show that under his conditions glucose is metabolised anaerobically and a significant quantity of lactic acid accumulates, indicating that the major function of glucose consumption is the provision of cellular energy. From studies on the accumulation of ammonium ions it seems likely that a significant amount of glutamine is deaminated and used as the energy source for oxidative metabolism in the tri carboxylic acid cycle. Although lactate itself is not inhibitory at the concentrations that typically accumulate in culture, the concentration of ammonium ions which accumulate has been shown to be toxic when added to fresh medium, and may be a major growth limiting factor [24].

Despite the extensive work conducted on medium development a formulation close to that originally described by Macpherson and Stoker [2] and supplemented with a quantity of peptone and a proportion of animal serum is generally preferred for commercial production. Although much has been written about the potential economies in medium formulation, the raw material costs for medium manufacture are generally low in relation to total manufacturing costs and, while the serum and peptones are the most expensive components, the defined alternatives prescribed are frequently even more expensive.

Development of cell sub-lines which grow in lower protein medium may result in alterations in other characteristics (see Sect. 2) including possibly the loss of susceptibility to virus. Hence each new formulation must be investigated critically as a cost effective medium for cell culture, without reducing the yield of final product.

Providing good quality serum is used for cell culture its concentration may be reduced below the 10% originally stipulated. Significant batch to batch variation can occur, however, and at levels of incorporation below about 5% serum, a diminishing proportion of batches will prove to be satisfactory. Furthermore, poor cell growth yields can have a greater effect on the efficiency and hence the cost effectiveness of the manufacturing operation than can marginally increased raw material costs. Since reliable and reproducible cell growth is a cornerstone in effective vaccine production, manufacturers may prefer to use a richer base medium adequately supplemented with peptones and serum in order to achieve this objective.

Although the nutritional requirement for animal serum may be met by substitution of other defined components the presence of serum seems to confer a protective effect which is important in defending cells against the shear forces developed by agitation and aeration in industrial culture systems. This aspect is discussed further in Sects. 4.3 and 4.4 but for this reason Kilburn [25,26] and later Radlett et al. [16] incorporated the pluronic polyol F68 into their culture medium, even when serum was also included in the formulation. More recent observations have shown, however, that providing other environmental factors are optimised the presence of protective agents other than a proportion of animal serum are not essential for industrial scale culture [15].

4 The Physical Environment

4.1 Temperature

Cell growth rate and yield are dependent on culture temperature and for maximum growth this must be closely controlled, usually around 35 °C [14]. A common method for controlling the temperature of large culture vessels is still by circulating water at incubation temperature through the jacket of the culture vessel. Telling and Elsworth [27] described a two step control action for mammalian cell culture using electric heaters and a water film cooler on vessels up to 100 litres working volume, and in our own laboratories we have extended this principle to the control of vessels up to 2500 litres in volume using steam heating directly to the vessel jacket and with no cooling facility. This indicates clearly that any medium degradation or cell death caused by local overheating is not of practical significance.

4.2 pH

It has been demonstrated many times that for maximum growth rate and cell production pH must be maintained within close limits, usually between pH 7.2–7.4 [28].

The common practice in tissue culture to regulate pH by adjusting the concentration of carbon dioxide in association with the bicarbonate buffer present in the medium has limitations at the industrial scale. As cell growth proceeds, the concentration of CO_2 required varies, and at high cell densities effective control becomes difficult [8]. The control of culture pH by automatic addition of acid or alkali is now well established and since cellular metabolism tends always to produce an acid drift we find it sufficient to use alkali only for correction. Although some workers have reported the importance of CO_2 *per se* in cell metabolism [29], we have never found its addition necessary for cell growth and the normal level following inoculation may well be sufficient for this purpose [25].

4.3 Oxygen

The importance of oxygen in the growth and metabolism of BHK cells was studied by Radlett et al. [30] who showed that under controlled conditions optimum yields were obtained when dissolved oxygen tension (pO_2) was not allowed to fall below 80 mm/Hg.

Kilburn and Webb [25] demonstrated that mammalian cells could be damaged by excessive sparging, and since that time Handa et al. [31] studied the effects of bubble diameter and superficial gas velocities on cells growing in bubble columns. They demonstrated that small bubbles are more damaging than larger ones and that increasing superficial gas velocities results in reduced viability. In optimising culture conditions for mammalian cells it is thus necessary to provide an adequate supply of oxygen without the damaging effects of excessive sparging. One obvious method of achieving this objective would be to measure and control the supply of air through the use of a steam sterilisable oxygen electrode situated within the culture. However, the reliability, stability and complexity of commercially available oxygen electrodes have deterred some manufacturers from committing large industrial cultures to such probes.

The oxygen demand for actively growing BHK cells has been estimated at a maximum value of 0.2 mM per 10^9 cells per h [30] and relating this data to information on the oxygen solution rates achieved from sparged aeration in individual culture vessels it is possible to calculate an air flow rate which satisfies the oxygen requirements of the culture without damaging the cells. Table 2 illustrates this method as applied to a range of geometrically similar culture vessels ranging in size from 30 to 2,500 litres. It is interesting to note that cell yields obtained using this technique were comparable to those obtained when pO_2 was not allowed to fall below a preset value [30].

Table 2. Calculation of operating conditions for BHK cell culture vessels

Agitator diameter	D/4
Agitator speed	150 rpm
Maximum oxygen demand	0.2 mM per 10^9 cells per h
Maximum demand from peak density of 5×10^6 cells per ml	1.0 mM l^{-1} h^{-1}
Air flow rate giving required solution rate	0.01 vvm
Sparged air flow rates giving 0.01 vvm	

Vessel size (L)	Air flow (L min^{-1})
30	$\simeq$0.5
300	3.0
1000	10.0
2500	25.0

An additional method for regulating the oxygen regime in the culture relies on using a Redox electrode to activate a control valve in the sparged air supply, such that air is blown into the culture fluid whenever Redox potential falls below a preset value. Although the relationship between Redox potential and dissolved oxygen tension is complex and will vary throughout the culture, depending on pH and medium composition, the use of Redox controller combined with regulating air flow rates by the method described above provides a simple and effective mechanism which permits adequate oxygenation without cellular damage.

4.4 Agitation

Mammalian cells may be irreparably damaged if agitated too vigorously, but different cell types vary in their susceptibility to mechanical damage. It has long been established that BHK cells can be grown satisfactorily in culture vessels designed originally for the growth of bacteria and agitated with a single turbine, if operated at relatively low speeds [27)]. In fully baffled systems speeds in the range 100–300 rpm have usually been found satisfactory to ensure adequate mixing, cell dispersion and oxygen solution without significant cell damage [32)].

A key problem in the maintenance of sterility is the reliability of the agitator shaft seal. The double radial mechanical seal used by Telling and Elsworth [27)] has proved satisfactory over the years but at the relatively low agitator speeds used for mammalian cell culture magnetic drive systems offer an attractive alternative. The system developed by Cameron and Godfrey [33)] but modified to encapsulate the internal magnet has proved entirely satisfactory for agitation of BHK suspension cell cultures in volumes up to 4000 litres.

5 The Manufacturing Process

A schematic outline of the production processes used for the manufacture of Foot and Mouth Disease Vaccines from BHK suspension cells is given in Fig. 1 [17)].

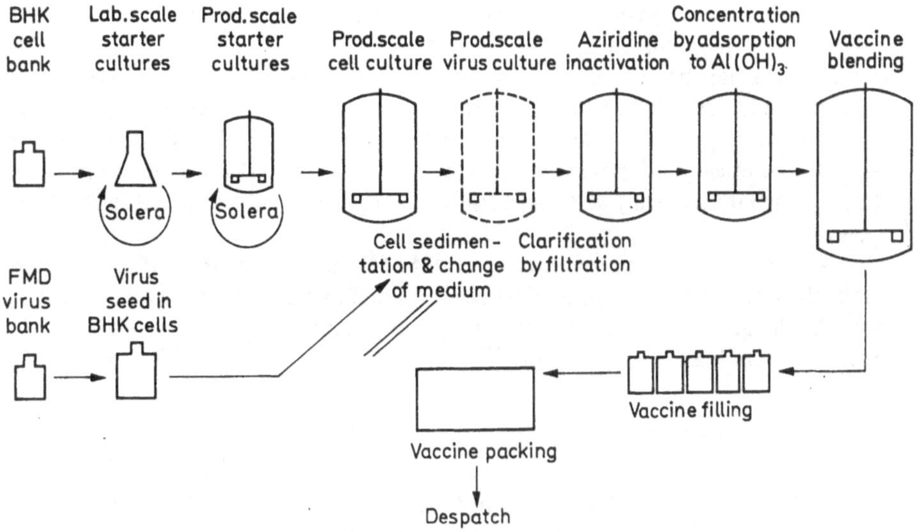

Fig. 1. Schematic outline of process for production of FMD vaccine in BHK suspension cells

5.1 Cell Growth

Banks of BHK 21 suspension cells tested for suitability for the process and freedom from adventitious agents are stored over liquid nitrogen and revived at approximately 10 week intervals into laboratory scale starter cultures. Cells are then grown through one or more intermediate vessels up to the full production scale, by means of a series of batch cultures in which surplus culture is removed and replenished with fresh medium. This is the so called "Solera" method described by Pirt [34]. Sufficient cells grown at the full production scale are sedimented to allow resuspension in fresh media prior to infection with virus.

5.2 Virus Production

Samples of FMD virus shown to have the required serological characteristics are tested for satisfactory antigen productivity in cell culture and for freedom from adventitious agents. Once adapted to growth in BHK cells suitably tested material constituting the master virus seed is stored in aliquots at low temperature and revived as required. The revived aliquot is propagated in BHK cells to produce a volume sufficient for seeding an industrial scale culture and used to infect the cells resuspended in the antigen production vessel. At the end of a one or two propagation stage this material is harvested.

Selection of appropriate virus strains is an essential and critical part of the whole manufacturing process. Seven distinct serotypes of Foot and Mouth Disease Virus are recognised, all of which are immunologically quite distinct in that vaccine produced from one type does not protect against another. Moreover, within each type there are many different subtypes and strains, with varying degrees of cross reactivity and protection [35]. The methods used to assess the serological suitability of strains for vaccine production have been described by Rweyemamu [36], but it is important to recognise that in order to be satisfactory for vaccine production a strain must not only be serologically appropriate for use in the geographical area for which vaccine is intended, but must yield productively in whatever culture system is involved. This is true for all manufacturing systems, however, and is by no means limited to the BHK process.

Coopers maintains a representative bank of FMD virus strains from all the infected areas of the world which is continually updated, and has pioneered methods of vaccine strain identification as a crucial component in the supply of vaccine appropriate for specific epidemiological requirements [36].

5.3 Antigen Clarification

Foot and Mouth Disease virus is lytic in BHK cells and is released directly into the culture fluid. At the end of the cultivation stage any remaining whole cells and cellular debris must be removed from the culture fluid. This is usually achieved either by centrifugation using a sterilisable continuous flow centrifuge or by filtration. In our hands, filtration is a less complex and more easily controlled alternative, for which the suitable process equipment is easily sterilised by steam in situ. One method of effectively removing cellular debris consists of recirculating culture fluid through a filter bed of diatomaceous earth, prior to terminal filtration using cartridge filters [17].

5.4 Inactivation

Of paramount importance in the production of inactivated vaccines against agents as infectious as FMD is the innocuity of the final product. It has been reported [37,38] that a high proportion of FMD outbreaks in Europe can be attributed to improperly inactivated vaccine and the adoption of inactivating agents which give a truly first order rate of reaction is of fundamental importance. As earlier demonstrated with polio virus, formaldehyde has been clearly shown to produce a "tailing effect" with FMDV [39,40] leaving an infectious residue in the harvest which may persist in the final vaccine. Despite the demonstrable difficulties, formaldehyde is still used by some manufacturers as an inactivating agent and Barteling [41] has recently described methods aimed at establishing a totally safe procedure for its use.

Aziridine inactivants are now widely recommended as the preferred alternative to formaldehyde, because, providing inactivation is carried out on carefully clarified culture harvests and in properly designed vessels, the rate of reaction follows first order kinetics [42].

Wellcome pioneered the industrial use of aziridine inactivants because they offered these advantages and the group has used them exclusively since 1965. The procedure developed involves adding two full doses of inactivant each sufficient to inactivate completely the harvest, at 24 h intervals. The procedure is controlled by sampling each production lot at regular intervals during the first inactivation cycle and testing for residual infectivity, from which an inactivation rate constant can be calculated.

In addition, separate samples are taken after incubation in the presence of one and two additions of aziridine and tested for innocuity in accordance with the requirements of the European Pharmacopoiea [17]. The double dose regime provides a high degree of product security and is controlled by the statutory innocuity procedures. In addition, calculation of the inactivation rate constant for each virus lot permits the identification and rejection of any atypical production material prior to vaccine formulation.

These procedures provide a very safe and well controlled production regime, and in twenty years of production, with a current level of vaccine offtake in the region of 350 million monovalent equivalent doses of vaccine per annum, there has never been a single confirmed instance of disease attributed to the presence of live virus in the vaccines manufactured in accordance with this procedure.

5.5 Antigen Concentration

Polyvalent vaccines against Foot and Mouth Disease are frequently required and depending on the geographical and epidemiological situation four to six different serotypes may be required in one vaccine blend. In order to produce satisfactory immunity a combined amount of antigen in excess of the normal vaccine dose volume is needed and thus some form of concentration is required. In addition, antigen concentration simplifies intermediate storage and handling and allows more cost effective utilisation of space and facilities. Concentration factors of 10–15-fold can be readily achieved by adsorption onto aluminium hydroxide but the requirement for strategic antigen banks held indefinitely over liquid nitrogen has encouraged the development of ultraconcentration techniques using polyethylene glycol [43,44], polyethylene oxide [45,46] or ultrafiltration [47].

The validity of these techniques for the concentration of BHK-produced antigen has been recently recognised by the establishment of the Seven Nation Strategic Reserve held at the U.K. Animal Virus Research Institute, and 50% of the antigen for this reserve has been provided from ultraconcentration BHK produced antigens with very satisfactory potency results.

5.6 Vaccine Formulation

Vaccines are formulated by incorporating adjuvants, preservative and diluent with an appropriate blend of suitable antigens. Various adjuvant systems have been used, but conventionally aluminium hydroxide and saponin are incorporated into aqueous formulations for cattle, sheep and goats. BHK-produced antigens may also be incorporated into oil adjuvanted emulsions which have proved very effective for protecting pigs against Foot and Mouth Disease. More recently similar vaccines have been registered for use in cattle in South America.

For both aqueous and oil vaccines the concentrated antigens of strains appropriate for the disease in the area for which the vaccine is intended are blended together at a payload sufficient to provide a high level of protection in the statutory potency test. In common with other methods of manufacture the payload of antigen required to confer immunity varies from virus strain to virus strain, and a higher payload is required where field and vaccine viruses are not *identical*. The latter situation may occur typically when vaccine is needed urgently to combat an outbreak involving a new serological virus variant. Modern analytical techniques such as the modified procedure for the measurement of antigen mass [48,49] and the application of ELISA [50,51] have greatly improved our understanding of the antigen payloads necessary for the formulation of effective vaccines and in most cases incorporation levels in the range 1–10 µg per dose prove satisfactory [52].

6 The Facilities

6.1 Culture Equipment

Experience has shown that with minor modification the vessels designed originally for bacterial culture are equally satisfactory for the production of viral antigen. The main differences arise in speed of agitation and the resultant option of driving the shaft through a pair of ceramic magnets rather than by utilising a conventional shaft seal.

Critical attention must be paid to detailed design and construction in order to ensure all equipment including valves and pipe layouts are self draining and crevice free for effective sterilisation by steam in situ. Provided that these aspects are catered for, complex plant layouts can be effectively sterilised and operated on a routine basis without difficulty. Because of the very nature of the manufacturing process it is frequently necessary to clean and sterilise part of the plant in isolation from the rest, and suitable valve arrangements must be provided to permit this. Suitable facilities for making additions and taking samples under aseptic conditions must be provided and this can be accomplished either by means of a portable laminar flow work station, or a suitably valved connecting assembly which permits in situ sterilisation without breaking the joint.

Under these conditions the entire manufacturing operation following the seed production stages can be performed in a completely enclosed system. Such plants have now been operated for many years without the need for sterile filtration of air coming into the room. Given good design, careful maintenance and skilled operators, losses due to contamination can be reduced to an insignificant level. In a recent survey of contamination in our laboratories losses of 2–3% (20 out of 740 cultures) for the cell growth stage and 1% (4 out of 464) for the virus production phase have been reported [17].

6.2 Production of Water

Traditionally, water for tissue culture has been produced by single or double distillation and this is the only method currently recognised by the European Pharmacopoeia for the production of water directly or indirectly intended for injection. From the purely technical standpoint the manufacture of suitable water will depend both on the purification process and the quality of the feed water. Experience gained in operating plants in various parts of the world has shown that providing the quality of the feed water is suitable, single deionisation is adequate to produce media which will support the growth of tissue culture. Of greater practical significance has been the development of ultrafiltration techniques for water purification (reverse osmosis). Because of their demonstrated efficacy in removing pyrogens and other toxic contaminants of feed water, ultrafiltration membranes have been accepted in the U.S.A. and Sweden as a means of producing water for injection, and the low energy running costs, minimal maintenance and moderate capital investment requirements compared to distillation makes this a highly attractive alternative technique for the purification of water where feedwater quality and legislative requirements permit. Our company has used such a system in its Brazilian plant for the past two years, without detecting evidence of pyrogenic activity downstream of the unit, and with highly satisfactory results in its culture systems. The savings against the use of oil as an energy source for distillation resulted in effective payback of capital investment in under one year.

6.3 Sterilisation of Media

Because tissue culture media contain heat labile components they are generally sterilised by filtration and for large scale operations filtration of completed media is generally considered more convenient and economic. Initially cellulose/asbestos depth filters were used [53] but the introduction of steam sterilisable cartridge filters with the characteristics of small pore size and high volume capacity has made possible the development of effective filtration techniques which obviate the need for asbestos-containing filters [54].

Although cartridge elements are expensive, significant economies in labour and filter turnround can be achieved which combined with reduced wastage (due to the high void volume hold up in traditional filter presses) can result in real cost benefits. Optimum savings will, however, depend on adequate prefiltration of serum-containing media to maximise cartridge life.

6.4 Sterilisation of Air

Animal cells require relatively low quantities of oxygen even at the peak densities normally achieved in free suspension culture, and in carefully designed equipment such requirements are easily met from sparged air at modest flow rates. Sterilisation of supply air is therefore simply accomplished by means of modern cartridge filters [55] and in the interests of containment and the maintenance of sterility it is sound practice to filter also the effluent air by means of similar units. These are best piped to permit independent steam sterilisation and should be carefully sited to minimise the chances of medium reaching the cartridges either as a result of foaming or overpressure in the culture vessel.

7 Quality Control

Quality control commences with the validation of all the raw materials used in vaccine production, continues with in-process testing during manufacturing and terminates with the evaluation of the final vaccine. The responsibility for quality control up to the completion of vaccine production normally rests with the manufacturer while the testing of the final bottled vaccine is usually vested in an independent control laboratory which assesses the product according to local pharmacopoeial requirements.

7.1 Control of Raw Materials

Control of raw materials embraces testing of process water, components of tissue culture media, filter media, inactivating agents, and all additives. Materials are usually tested for cell growth potential and lack of toxicity in tissue culture.

7.2 In-process Controls

In-process controls include routine tests and check procedures to ensure that the sequential steps of the production process are correctly executed, that equipment is functioning properly and that process performance is within normal limits. This includes validation of plant in respect to safety, efficiency and performance and a battery of control tests performed on each individual production lot. Currently such routine testing might be expected to cover estimation of cell density, testing of cell substrates for susceptibility and productivity, identity testing of virus seeds, assessment of antigen production (often by several different methods) and integrity testing by polyacrylamide gel electrophoresis. Comparison of seed virus stocks and production harvests may be made using "finger printing" techniques [56] and sterility is monitored throughout the production process. The manufacture of oil adjuvanted vaccine is accompanied by additional tests to confirm emulsion type and stability, viscosity and conductivity. These test procedures are carried out and meticulously documented for each individual antigen and every vaccine batch and these records together form a complete manufacturing history following the guidelines established by the British Guide to Good Manufacturing Practice [57].

7.3 Tests on Vaccine

Fully formulated vaccine is subjected to tests for sterility, abnormal toxicity and potency. In most cases this involves testing both "in vitro" and in animals. These tests may involve laboratory animals, such as guinea pigs, or the target species, (usually cattle and pigs) and may be by direct challenge or by the indirect assessment of serum neutralising antibody levels. Testing in the target species is the ultimate criterion of potency, but the preferred method is by antibody assay where a correlation has been established between serum neutralisation titre and protection against challenge. This indirect method is advantageous on humanitarian, disease security and economic grounds since it does not involve causing disease in the test animals.

While acceptance criteria have been stipulated in a number of national pharmacopoeias there is as yet no universally accepted potency test method or pass level for FMD vaccine. However, examples of widely used methods include: the percentage protection test in cattle or pigs using undiluted vaccine with a pass level of at least

70% protected on challenge; the determination of the number of fifty percent protective doses per vaccine dose using dilutions of vaccine in mock vaccine (the so-called PD_{50} method) or in carbonate buffer (the so-called PB_{50} method) with pass levels of at least 6.0 PD_{50} or at least 3.0 PB_{50} per vaccine dose. The K Index Method derived from the difference between challenge virus titre in groups of vaccinated and non-vaccinated cattle and the C Index Method similarly derived in guinea pigs. K Index and C Index pass levels are usually at least 1.2 or 2.0 respectively.

Limitations of biological testing include high costs, which normally preclude the use of large numbers of cattle and pigs, and consequently wide variation in results. The subject is complex and beyond the scope of this review but more complete discussion is available [58]. Nevertheless, the methods listed above have been adequate for the testing of millions of doses of vaccine which have given satisfactory control of the disease in the field. Vaccines tested by such methods normally exceed the minimum acceptance levels by a wide margin and have been associated with excellent control of the disease in many areas of the world and even eradication in certain countries.

Of increasing concern in the manufacture of all vaccines is the low number of animals which may develop allergenic reactions. Although this may occur with vaccines from any source, and the numbers of reacting animals are usually very small in relation to the number of vaccine doses used, minimising such problems remains a critical objective in quality control. Derivatives from BHK cells have been investigated as a possible source of allergenic material [59] but so far tests have given equivocal results. For example field surveys in West Germany covering several million cattle on each occasion showed that FMD vaccines produced from cattle tongue epithelium produced more reactions than BHK vaccines in the 1968/69 and 1970 campaigns [60,61] but fewer reactions in the earlier 1967/68 campaign [62].

Not only must modern vaccines be potent and free from side reactions, they must also meet the requirements of stringent sterility test procedures. As with other tests on final product, the statutory requirements may differ between countries but in most cases test procedures similar to those described in the European Pharmacopoeia must be performed using batches of media which in themselves have been shown to support the growth of low numbers from a range of test organisms.

The success rate in final quality control tests with vaccines manufactured using the BHK process is usually high and typically our group in 1983 submitted 160 batches of such vaccine comprising 350 million monovalent equivalent doses to independent control authorities in eight different countries and achieved a success rate of 97% [17]. Such rates of success however are not achieved by chance, but by vigilant monitoring and testing at all stages of the manufacturing process, such that only materials meeting the most stringent quality requirements are incorporated into the final product.

8 Conclusions

The BHK suspension cell process has been used commercially for the production of Foot and Mouth Disease vaccine for over 20 years. In that time many millions of doses have been produced, and the use of these vaccines has been specifically associated with the successful control of FMD in countries such as Chile, Indonesia, Uruguay and the Philippines.

Microbial contamination remains the biggest single risk to the process but with properly designed equipment and well trained operators production plant can be managed with a high degree of efficiency and reliability. Assessment of quality remains essentially subjective because of the very nature of disease control programmes, and the fact that many different quality standards are used in the different countries in which vaccines are manufactured and used.

Undoubtedly some problems remain, and these include matching production and challenge strains of virus, and adapting new virus strains to grow in culture [5]. Both of these problems may occur with all conventional manufacturing methods and are of concern to both manufacturers and control authorities alike. New developments in nucleotide sequencing and peptide synthesis are improving our understanding of the fundamental differences between virus strains and hence their significance for vaccine manufacture, thus offering a real hope of now standardising some of the variables which all manufacturers and control authorities face. Although these new techniques provide great promise there is, as yet, no single in-process parameter which can be used with confidence for all strains of virus to predict the performance of antigens when formulated into vaccine and tested in animals. Although much variation in cell substrates is currently apparent, cloning and hybridisation techniques may yet provide the tools for the development of new stable highly productive cell substrates with a broad spectrum of virus susceptibility.

The possibilities for peptide synthesis clearly extend beyond the analysis of process variables, and a totally synthetic vaccine against foot and mouth disease remains a realistic objective [63]. Nevertheless, the hamster kidney production process has served well in the control of FMD and the cost effectiveness of the new technologies must in this instance be measured against existing production methods. Furthermore the experience gained from operating a tissue culture process at this scale has directly contributed to the provision of suitable technology for the production of rabies vaccine [64] and interferons [65].

9 Acknowledgements

I would like to thank Mrs. M. Draper and Dr. A. Garland for their helpful criticism and contributions and Mrs. A. Wright for her careful preparation of the manuscript.

10 References

1. Frenkel, H. S.: Amer. J. Vet. Res. *11* (1950)
2. Macpherson, I. A., Stoker, M. G. P.: Virology *16*, 147 (1962)
3. Mowat, G. N., Chapman, W. G.: Nature (Lond.) *194*, 253 (1962)
4. Capstick, P. B., Chapman, W. G., Stewart, D. L.: ibid. *195*, 1163 (1962)
5. Capstick, P. B., Garland, A. J., Masters, R. C., Chapman, W. G.: Exp. Cell Res. *44*, 119 (1966)
6. Pay, T. W. F. P., Boge, A., Ménard, F. J. R. R., Radlett, P. J.: Develop. Biol. Standard *60*, 171 (1985)
7. Zoletto, R., Gagliardi, G.: Paper presented at the 16th Venetian Regional Conf. of the Italian Association of Hygiene and Public Health (1968)
8. Telling, R. C., Radlett, P. J.: Paper presented at the Meeting of the Standing Tech. Comm., European Comm. for the Control of Foot and Mouth Disease, Brescia (1969)

9. Ubertini, B., Nardelli, L., Dal Prato, G., Panina, G., Barci, S.: Zbl. Vet. Med. *14*, 432 (1967)
10. Handa, A.: Dept. of Chemical Engineering, Univ. of Birmingham. Personal communication
11. Spier, R. E., Clarke, J. B., Preston, K. J., Mowat, G. N.: Paper No. 305 XVième Conf. Comm. Perm. Fievre Aphteuse de l'O.I.E., Paris 1978
12. Draper, M. E.: Coopers Animal Health Ltd. Pirbright. Personal communication
13. Clarke, J. E., Spier, R. E.: Ariv. Virology *63*, 1 (1980)
14. Capstick, P. B.: Proc. Royal Soc. Med. *56*, 1062 (1963)
15. Panina, G. F.: Paper presented at the Meeting of the Research Group, European Comm. for the Control of Foot and Mouth Disease, Brescia 1985
16. Radlett, P. J., Telling, R. C., Stone, C. J., Whiteside, J. P.: Applied Microbiology *22*, 534 (1971)
17. Radlett, P. J., Pay, T. W. F., Garland, A. J. M.: Dev. Biol. Standard *60*, 163 (1985)
18. Keay, L.: Biotech. Bioeng. *17*, 745 (1965)
19. Mizrahi, A., Avihoo, A.: J. Biol. Standard *4*, 51 (1983)
20. Mizrahi, A.: Biotech. Bioeng. *19*, 1557 (1977)
21. Arathoon, W. R., Telling, R. C.: Dev. Biol. Standard *50*, 145 (1982)
22. Butler, M.: ibid. *61*, 269 (1985)
23. Butler, M.: Paper presented at the Meeting on Process Possibilities for Plant and Animal Cell Culture, Manchester 1986
24. Butler, M., Spier, R. E.: J. Biotech. *1*, 187 (1984)
25. Kilburn, D. G., Webb, F. C.: Biotech. Bioeng. *10*, 801 (1968)
26. Kilburn, D. G., Lilly, M. D., Self, D. A., Webb, F. C.: J. Cell Sci. *4*, 25 (1969)
27. Telling, R. C., Elsworth, R.: Biotech. Bioeng. *7*, 417 (1965)
28. Telling, R. C., Stone, C. J.: ibid. *6*, 147 (1964)
29. Geyer, R. P., Change, R. S.: Arch. Biophys. *73*, 500 (1958)
30. Radlett, P. J., Telling, R. C., Whiteside, J. P., Maskell, M. A.: Biotech. Bioeng. *14*, 437 (1972)
31. Handa, A., Emery, A. N., Spier, R. E.: Develop. Biol. Standard (In press) Vienna (1985)
32. Telling, R. C., Radlett, P. J.: Advances in Appl. Microbiol. *13*, 91 (1970)
33. Cameron, J., Godfrey, E. I.: J. Appl. Bacteriol. *31*, 405 (1968)
34. Pirt, S. J., Callow, D. S.: Exp. Cells. Res. *33*, 413 (1964)
35. Pay, T. W. F.: Revue Scientifique et Technique de l'Office Int. des Epizooties *2*, 701 (1983)
36. Rweyemamu, M. M.: J. Biol. Stand. *12*, 323 (1984)
37. Council Directive 84/C248/08 amending Directive 64/432/EEC. Official Journal of the European Communities (1984)
38. Strohmaier, K., Bohm, H. O.: Tierärztl. Umschau *34*, 949 (1984)
39. Graves, J. H.: Am. J. Vet. Res. *24*, 1131 (1963)
40. Fontaine, J., Farne, H., Fargeaud, D., Roulet, C., Dupasquier, M.: Bull. Off. Int. Epizoot. *81*, 1089 (1974)
41. Barteling, S. J., Woortmeyer, R.: Archiv. Virology *80*, 1089 (1974)
42. Pay, T. W. F., Telling, R. C., Kitchener, B. L., Southern, J.: Paper presented at the meeting of the Research Group, European Comm. for the Control of Foot and Mouth Disease, Tubingen 1971
43. Lei, J. C.: Bull. Off. Int. Epiz. *81*, 1169 (1974)
44. Lei, J. C.: Paper presented at the Research Group, European Comm. for the control of Foot and Mouth Disease Tubingen 1971
45. Adamowicz, Ph., Legrand, B., Guerche, J., Prunet, P.: Bull. Off. Int. Epiz. *81*, 1125 (1974)
46. Duchesne, M., Guerche, J., Legrand, B., Proteau, M., Colson, X.: Dev. Biol. Stand. *50*, 249 (1982)
47. Moore, D. M., Morgan, D. O.: Report of the Research Group, European Comm. for the Control of Foot and Mouth Disease, Lindholm, Denmark 1979
48. Barteling, S. J., Meloen, R. H.: Archiv für Virusforschung *45*, 362 (1974)
49. Rweyemamu, M. M.: Proc. 1st Int. Conf. on Impact of Viral Diseases on Dev. Latin American Countries, Rio de Janeiro, 437 (1982)
50. Ouldridge, E. O., Barnett, P. V., Hingley, P. J., Rweyemamu, M. M.: J. Biol. Stand. *12*, 339 (1984)
51. Ouldridge, E. O., Barnett, P. V., Hingley, P. J., Rweyemamu, M. M.: ibid. *12*, 367 (1984)
52. Pay, T. W. F., Hingley, P. J., Radlett, P. J., Black, L., O'Reilly, K. J.: Report of the Research Group, European Comm. for the Control of Foot and Mouth Disease. Lelystad, Appendix IX, 52 (1983)
53. Telling, R. C., Stone, C. G., Maskell, M. A.: Biotech. Bioeng. *8*, 153 (1966)

54. Ball, G. C.: Clarification and Sterilisation, in Animal Cell Biotechnology (eds. Spier, R. E., Griffiths, B.) Vol. 2, Academic Press 1985
55. Conway, R. S.: Paper presented at the American Chemical Society, Kansas City (1982)
56. La Torre, J. L., Underwood, B. O., Lebendiker, M., Gorman, B. M., Brown, F.: Infect. and Immun. *36*, 142 (1982)
57. Guide to Good Manufacturing Practice (Ed. Sharp, J. R.) Published by Her Majesty's Stationery Office 1983
58. Pay, T. W. F., Parker, M. J.: Paper presented at the Internat. Symp. on FMD, Lyon (October 1976)
59. Black, L.: Vet. Record. *100*, 195 (1977)
60. Lorenz, R. J., Straub, O. L.: Zentbl. Bakt. Parasitkde. Abt. 1. Orig. A. *223*, 1 (1971)
61. Lorenz, R. J., Straub, O. L.: ibid. A. *223*, 1 (1973)
62. Mayr, A., Mussgay, M.: Report of Research Group European Comm. for the Control of Foot and Mouth Disease, Brescia, Italy 1969
63. Brown, F.: 4th Meeting of European Group of Molecular Biology of Picornaviruses, Seillac, France (1985)
64. Pay, T. W. F. et al.: Dev. Biol. Stand. *60*, 171 (1985)
65. Pullen, K. F. et al.: Dev. Biol. Stand. *60*, 175 (1985)

Production of Tissue Plasminogen Activators from Animal Cells

J. Bryan Griffiths[1] and Asgar Electricwala[2]

[1] Vaccine Research & Production Laboratory; [2] Microbial Technology Laboratory, PHLS Centre for Applied Microbiology and Research, Porton Down, Salisbury, Wiltshire, England

Plasminogen activators catalyse the conversion of plasminogen to the active fibrinolytic enzyme, plasmin. They are widespread in the body and play an important role in the balance between fibrinolysis and coagulation. Although associated with many physiological events it is their potential therapeutic role as a specific thrombolytic agent against occlusive vascular disorders which is exciting most attention. Plasminogen activators have been isolated from cultured cells for many years but only in very small amounts. The discovery of a human melanoma cell line that secreted elevated amounts of a tissue-type plasminogen activator (t-PA) opened the way for large-scale purification and subsequent characterisation of this enzyme together with in vivo clinical trials. This review describes the development of the production of t-PA from various types of cultured cell and its present status as a drug undergoing clinical trials. Thus production characteristics from various cell lines, increase of yields by chemical inducers and by recombinant DNA technology, purification, characterisation of its biochemical properties and preliminary clinical results are summarised. The fact that occlusive disorders are one of the largest causes of death in the Western world underlines the importance of this cell product in health care.

1 Introduction

Plasminogen activators are serine-type proteases which catalyse the conversion of a precursor, plasminogen, to the active enzyme, plasmin. This latter enzyme plays an important role in fibrinolysis, i.e. proteolytic breakdown of fibrin clots, and is part of the body's haemostatic control to balance coagulation. There are two recognised classes of plasminogen activator, tissue type (t-PA) and urokinase type (u-Pa) based on differences in their immunological and biochemical properties [1,2,3]. u-Pa is generally regarded as being relatively non-specific in its action and some forms can activate both circulating and fibrin bound plasminogen indiscriminately, thereby giving rise to serious risk of haemorrhage. By contrast, t-PA is regarded as being far more specific in its mode of action, binding relatively strongly to fibrin clots and preferentially activating the plasminogen entrapped in blood clots, but sparing the circulating plasminogen and other blood coagulation factors. It has been shown that the tissue type activator is immunologically identical to the vascular activator. Plasminogen activators are very widespread in the body being found in all body fluids, (blood, lymph, urine, cerobrospinal fluid, etc.), many tissues and in a wide variety of tumours [4]. The richest sources of activators are the human uterus [1,5] (1 mg per 5 kg of uterus), hog ovary [6], pig heart [7] and the endothelial cells lining the blood vessels [8,9,10]. Although these latter cells only normally secrete small amounts of t-PA into the blood stream the concentration is increased many-fold in response to certain stimuli e.g. chemicals (such as adrenalin, vasopressin, histamine, bradykinins), the nervous system and various physiological states such as anoxia [11,12]. Plasminogen activators are associated with many physiological and pathological events [13] including thrombolysis [4], cell migration, (mononuclear phagocytes [14]) transformation [15] and metastasis [15], rheumatoid arthritis [15] and wound healing [1]. The association between tumour cells and high levels of fibrinolytic activity has been accepted as an indication of high protease secretion by these cells and thereby as the means by which they invade normal tissue, using the proteolytic enzymes to degrade connective tissue [16,17]. This difference between normal and neoplastic cells has also been found between normal and in vitro transformed cells and has been investigated as a basis for testing for malignancy. Under normal physiological conditions plasminogen activators are associated with all tissue remodelling in the body which occurs during embryonic tissue development and following ovulation [18] and mammary involution [19]. However, it is their therapeutic role as a specific thrombolytic agent which is causing most attention at the moment and which needs the development of large-scale production processes. The potential for t-PA is in the treatment of acute myocardial infarction, deep vein thrombosis, pulmonary embolism and cerebral stroke, together the commonest causes of death in developed countries.

2 Biochemical Properties

Most of the data on the biochemical and structural properties of t-PA have been obtained from Bowes melanoma t-PA due to its availability in sufficient yields [20]. Tissue plasminogen activator is a serine protease with a molecular weight of approximately 70,000 Da and composed of a single polypeptide chain containing 527 amino acids,

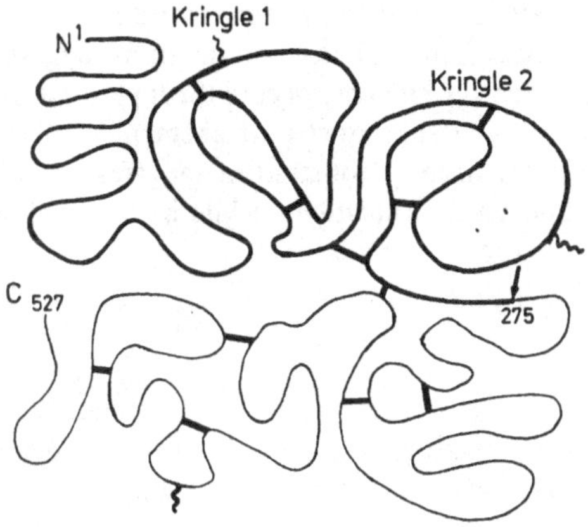

Fig. 1. Schematic diagram for the structure of human t-PA. — Arrow indicates the site of cleavage between Arg^{275}-Ile bond. ▬: disulphide bonds; ∿∿: glycosylation sites. N, C denote amino and carboxyl terminals respectively

with 35 cysteine residues and three potential N-glycosylation sites (Fig. 1). Like urokinase, it is synthesised within the cell as a single polypeptide chain and released as a one chain enzyme after cleavage of the leader sequence. Upon limited proteolytic action by plasmin in the extracellular medium, the molecule is cleaved between Arg 275 and Ile 276 peptide bond and converted to a two chain activator linked by an inter-chain disulphide bond [7]. The two chain activator consists of a heavy chain from the amino-terminal part of the molecule and a light chain from the carboxyl-terminal region of the protein. The heavy chain contains two looped structures called "kringles" of 82 amino acids that share a high degree of homology with the kringles of plasminogen, prothrombin and urokinase. Although the kringles of t-PA have not been studied in detail, there is evidence to suggest that the kringles contain lysine binding sites which anchor the molecule to the fibrin network of the blood clots [21]. The heavy chain of t-PA also contains a finger domain of 43 amino acids at the N-terminal part of the molecule [21], followed by a cysteine rich domain from residues 44–91 which is relatively homologous with the epidermal growth factor. Indications are that the N-terminal finger domain, which is absent in urokinase, is responsible for the high fibrin affinity of t-PA. The light chain contains the presumed catalytic site and is formed by His 322, Asp 371 and Ser 478 residues. The amino acid sequences surrounding these residues are highly homologous to the active sites of other serine proteases such as trypsin, thrombin and plasmin [22].

The one chain and two chain forms of tissue activator have virtually the same fibrinolytic activity in a purified clot lysis assay, but have different amidolytic activities towards low molecular synthetic substrates [23,24]. It has been observed that on the fibrin surface, the one chain form is quickly converted to the two chain form and therefore it has been suggested that physiological fibrinolysis induced with native one chain activator occurs via a two chain molecule.

Apart from one and two chain forms of the molecule, t-PA also exists in two different forms depending upon the position of glycosylation [25,26,27]; type I is glycosylated at position 118, 186 and 448 and has a molecular weight of 63,000 Da; type II is glycosylated at position 118 and 448 only and has a molecular weight of 60,000 Da. Thus, purified t-PA appears as a doublet on SDS-polyacrylamide gel electrophoresis. Both these glycosylated forms have similar enzyme activities.

t-PA is a relatively poor plasminogen activator in the absence of fibrin but its activation rate is stimulated several fold in the presence of fibrin. The kinetic analysis suggests that the activation, which obeys Michaelis-Menton kinetics, occurs through binding of an activator molecule to the fibrin surface and subsequent attachement of plasminogen. In physiological fibrinolysis, the low affinity constant in the presence of fibrin allows efficient plasminogen activation on the thrombus, while its high value in the absence of fibrin prevents efficient activation in blood [1].

3 Production by Cells

The widespread distribution of plasminogen activators throughout the body tissues is reflected by the many types of cells that have been found to secrete these enzymes. The first activator to be produced from cells in culture was the urokinase-type from human primary embryonic kidney cells [28,29] and typical production yields were 5–10 IU per 10^6 d^{-1} cells. However, this low yield may have been due to the fact that only a small proportion of the cells in culture, estimated at 5% [30], were producing plasminogen activator rather than a low yield from each cell. The procedure used was to grow the cells to confluency, change the medium to a peptone supplemented serum-free medium, and then harvest u-PA at intervals for 1 month. The range of cell types and species of derivation, and incidentally the vast effort put into studying cultured cell plasminogen activators, is illustrated in Table 1. Many of the references given in Table 1 do not specify which sort of activator was being produced and it must be assumed that many, if not most, were of the urokinase-type. Wilson and co-workers [31,32] studied over 94 human cell lines of both normal and neoplastic origin. In one paper [31] 56 lines were examined in an attempt to discover if there was a pattern between the cell type and type of activator produced. Although no strict demarcations could be made, it appears that cells derived from melanomas, certain other malignant neoplasms and embryonic tissues secrete a t-PA. Most normal adult tissues (apart from 2 exceptions) and the majority of the neoplastic lines secreted u-PA [31]. This assessment is further complicated by the fact that many cell types (and lines) secrete both u-PA and t-PA, examples include bovine endothelial cells [33], prostrate tissue [34], human leukemic cells [35] and human fibroblasts [16]. Physiological reasons for this dual production are unknown and may be related to different functions. The possibility that this phenomenon is due to population heterogeneity has been tested by cloning experiments and found to be unlikely [33].

Increased fibrinolytic activity is a characteristic property of cells transformed by viral and chemical agents and activator secretion has been investigated as a marker for tumour and transformed cells. The enhanced secretion of activators from many species (chicken, hamster, mouse and rat) on transformation by oncogenic viruses, such as SV40 and the Rous sarcoma virus, is well documented [17,36,37,38,39]. High levels of activator production have also been correlated to the expression of the malignant phenotype [37] and induction of carcinogenesis by tumour promoters, and is thought to enhance tumour invasiveness by causing destruction or alteration of normal tissue [16][17] thus allowing tumours to spread. The degree of enhancement between normal and transformed cells is in the region of ten-fold [17,39] but this differential disappears when normal cells are treated with various chemical stimulators (see

Table 1. Plasminogen activator production from cultured cells

Species/Cell type	Designation	Ref.
Human		
Diploid fibroblasts	W1-38, IMR-90	44,64)
	D549	66)
	HDC, various tissues	39,62–65)
	Rheumatoid Synovial	15)
Transformed fibroblasts	SV WI-38-VA13	41)
Tumour cells and lines	HeLa	99–101)
	Detroit 562	44)
	Melanoma, Bowes	13,20,39,46,102)
	Melanoma, Malme	103)
	Adenocarcinoma	66)
	Fibrosarcoma	104)
	Liposarcoma	39)
	Oesophageal carcinoma	65)
Epithelial cells	BEB	42,57,61)
	HBL 100	66)
	Mammary HME	16)
	Polymorph, leucocytes	105)
Lymphocytes		80)
Endothelial cells	Vascular	69,74)
Other species:		
Diploid fibroblasts	Chick	39,81,101,106)
	Hamster	39,101)
Primary cells	Chick, various	28)
	Chick, enbryo muscle	78)
	Rat	101)
Fibroblast cell line	Mouse 3T3	20,36,38)
	L cell	101)
	SV_{40} 3T3	36,40)
Epithelial cells	PK15 (pig kidney)	107)
	LLC-PK (pig kidney)	49,68)
	GPK (guinea pig keratocyte)	42,57,61)
	Mouse — teratoma	108)
	Mouse — teratocarcinoma	109)
	Mouse — macrophages	14,118)
	Mouse — mammary	37,66)
	Mouse — carcinoma	110)
	Mouse — melanomas	111)
	Mouse — tumour	111)
	Rat — hepatoma	101)
	Rat — ovarian granulosa	82,108)
Endothelial cells	Bovine carotid artery	70,71)
	Rabbit vena cava	72,112)

Table 3). Plasminogen activator is predominantly extracellular but significant intracellular levels have been found in some cell lines and thus care has to be taken in assessing the effect of stimulators, with respect to causing an actual production increase rather than a rapid secretion of intracellular activator. Usually the extracellular concentrations are at least 3-fold higher [40]. Other factors which could affect apparent

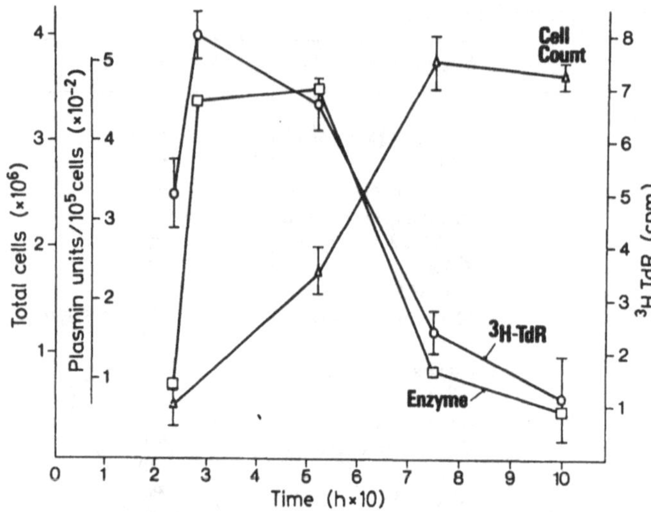

Fig. 2. The production of t-PA by epithelal cell line (GPK). At each time point the medium was replaced with serum-free medium and [^{3}H]-thymidine for 0.5 h. This medium was then assayed for t-PA and shows a correlation between t-PA production and growing, not stationary, cells. (Data supplied by P. A. Riley and S. Naish, University College Medical School, London)

yields are the presence of inhibitors either in the serum or secreted by the cell [41]. In many cell lines plasminogen activator production is growth-dependent with maximum expression during the growth phase and low expression when the cell sheet is confluent. This relationship has been demonstrated in normal fibroblasts from many species [39], 3T3 cells [36], and human and guinea pig epithelial cell lines [42] (Fig. 2). The main activator associated with malignancy is the u-PA [43,44] but an important exception is the melanoma. The Bowes melanoma cell line is very stable and produces large amounts of a t-PA that is very similar to that extracted from the human uterus [31,45]. The use of this cell line has enabled large enough quantities of t-PA to be produced. Thus, publications since 1979 have usually been able to specify the type of activator being produced.

3.1 Melanoma Cells

The importance of the Bowes melanoma cell line to the development of our knowledge of activators and to a possible therapeutic agent is two-fold. Firstly, it is a genuine tissue type activator very similar to that found in normal human tissue [20]. Secondly, its high level of production has provided sufficient material for scientific investigation. One of the richest sources was the human uterus which yields about 0.01 mg of purified enzyme per uterus. The Bowes melanoma line produces over 30 IU per 10^6 cells per day (0.1 mg L^{-1} of conditioned medium) [46], i.e. the equivalent of about 10 human uteruses. Despite this level of production, a manufacturing problem still exists for, based on a clinical dose in the region 7.5 mg [47], then 60–75 L of conditioned medium per dose are needed. The only way that an economical production schedule can be introduced would be as a result of genetic engineering, and the first steps towards this have already been taken [27,48–53].

The most successful production method for t-PA is by growing the Bowes melanoma cells on microcarriers [46]. Attempts to grow the cells in suspension with the same enzyme expression level were unsuccessful [46] so the method of growing cells attached to small (200 micron) spheres (microcarriers) which can be stirred in suspension cultures was adopted. The cells were grown on the collagen coated microbeads (Cytodex-3™) and attachment was very dependent upon a critically controlled pH and the presence of serum (0.5%) in the medium. Cell and enzyme yields were found to be identical to that obtained in conventional polystyrene culture dishes. The culture size was 40 L with 3–5 g L^{-1} Cytodex-3 (surface area of $14–23 \times 10^3$ cm^2 L^{-1}). Cells were inoculated at 1×10^5 cells per ml in a medium supplemented with 10% serum, grown until stationary and then resuspended in the production medium (0.5% serum). The count of approximately 1×10^6 cells per ml, and a daily harvest of about 15 IU ml^{-1} was maintained for 700 h before the culture had to be terminated.

This experimental process [46] demonstrated the feasibility of producing t-PA on a continuous basis from cells in culture. Initially clinical doses of 7.5 mg were used [47] but more recently doses of 80 mg. per patient are being given [54,55]. Thus the daily yield from a 40 L culture falls far short of a clinical dose and even the prospect of being able to scale-up this operation to 2000–4000 L bioreactors means a very expensive product. A favourable aspect of this process is that it is not a batch production but a semi-continuous process capable of running for about 1 month. However, the future of t-PA as a clinical product obviously will rely not on this product but on t-PA produced by recombinant DNA technology (see 3.5).

3.2 Epithelial Cells

A t-PA derived from normal epithelial cells in culture has been extensively studied and characterised [42,56–60]. The initial interest was to obtain large amounts of a t-PA from normal, as opposed to tumour, tissue for evaluation as a therapeutic agent. However, the characterisation studies of this epithelial t-PA have shown it

Table 2. Comparative yields of t-PA from melanoma, epithelial and recombinant DNA cell lines expressing melanoma t-PA

Cell line	Production rate		Ref.
	pg per cell per day	IU per 10^6 d^{-1} cells	
Bowes melanoma	0.25	15	20)
Bowes melanoma	0.31	18	48)
r Bowes melanoma	3.10	180	48)
GPK	0.38	3	58)
GPK (+ azacytidine)	1.26	10	58)
GPK (+ concanavalin A)	5.80	46	58)
r CHO	0.86	50	51)
r CHO (+ methotrexate)	8.61	500	51)

to possess some biochemical properties which are different to Bowes melanoma t-PA [56,119]. Epithelial cell t-PA differs from melanoma in that it has a lower specific activity (12,500 IU mg^{-1} compared to 90,000) and isoelectric point (4.7 compared to 7.8). Thrombolytic activity on human blood clots has been found to be similar for both types of t-PA.

The source of this epithelial t-PA is from guinea pig ear keratocytes (GPK cell line) and human breast epithelial cells (BEB line) [61]. These cells show great proliferative ability in culture and, although derived from normal origin, do not show the classical senescent symptoms at passage levels under 60. However, the enzyme yield does show a decline with culture age. The cells are very dependent upon serum for growth which poses a production problem as the highest expression of enzyme is during the growth phase [42] (Fig. 2) — and a low protein medium is required to aid the purification of the enzyme from the culture supernatant. The production process, which has been published [42,59] is based on high density microcarrier culture (Cytodex-3 carriers at > 12 g L^{-1}). A batch process is used in which cells are grown to 60 to 70% confluency, the medium changed to the production formulation (serum-free, plus 0.01% Tween-80 and stimulating agents), and the enzyme harvested approximately 60–80 h later. To offset the disadvantages of a batch process, compared to semi-continuous, the cells are cultured at a high cell density ($7–10 \times 10^6$ per ml) by means of sophisticated perfusion and aeration adaptations to standard cell reactors

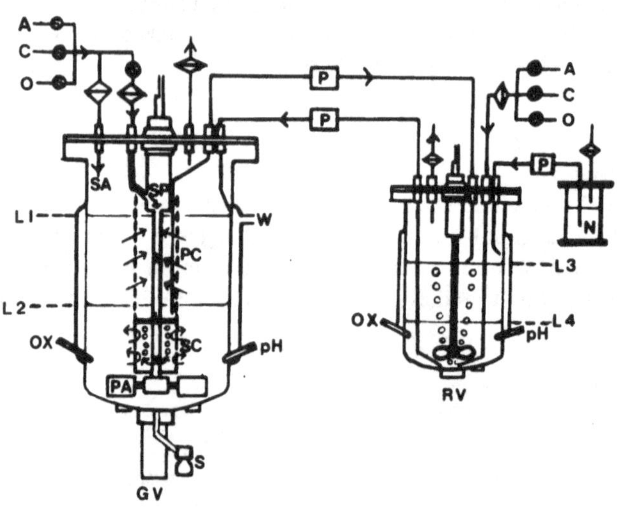

Fig. 3. Culture equipment used for the large-scale production of t-PA in perfused microcarrier culture of GPK cells. A — air; C — carbon dioxide; GV — growth vessel; Ll — maximum working volume; L2 — minimum working volume (product generation volume); L3 — growth phase volume (reservoir); L4 — product generation volume (reservoir); M — marine impeller; N — NaOH reservoir; O — oxygen; OX — oxygen sensor; P — peristaltic pump; pH — pH sensor; PC — perfusion compartment of filter; PA — stirring paddle; RV — reservoir vessel; Ⓢ — solenoid valve; SA — surface aeration; SP — sparging through stirrer shaft; SC — sparging compartment of filter; S — sampling port; W — waterjacket. Cells are grown on Cytodex microcarriers (12 g L^{-1}) for 72 h, medium changed to serum-free and volume reduced to increase enzyme concentration in supernatant. At cell counts above 7×10^6 per ml the perfusion rate is increased to 1 volume h^{-1} to maintain oxygen levels

(Fig. 3). The yields obtained are in the order of 5–10 mg of purified t-PA per litre of conditioned medium. The lower specific activity of epithelial t-PA (12,500 IU mg^{-1}) compared to melanoma t-PA (90,000 IU mg^{-1}) means that despite the very much higher protein yield in stimulated epithelial cell cultures productivity, in terms of units, is only similar to the native melanoma system when potentiated by various agents (Table 2). It must be emphasised that the epithelial t-PA is produced in a batch system so these high yields represent a single, or at most double, harvest and not a daily harvest over several weeks. Assuming a batch culture is over 7 days then a 5–10 mg (63,000–126,000 IU L^{-1}) yield has to be compared with 7 daily yields of 0.1–0.15 mg i.e. 0.7–1 mg (63–90,000 IU) of melanoma t-PA.

3.3 Fibroblast Cells

Many normal fibroblast cell lines secrete small amounts of t-PA or u-PA, or both. The wide range of human fibroblast cells [39,62,63,64,65] that have been found to secrete t-PA include the well established lines W1-38 and IMR-90 [64,66]. These cell types are also responsive to phenotypic stimulation to increase the enzyme secretion level by twenty-fold [64]. Initially, agents such as the prostaglandins [67], catecholamines [67] and retinoids [68] were used but the highest productivity (100–300 U per 10^6 cells per day) has been achieved by pre-coating the culture substrate with poly-D-lysine and using alternative sera to foetal calf, for example horse serum [64]. This is an adaptive process in that production returns to the normal low levels (1.0 U per 10^6 cells per day) if the cells are returned to non-permissive conditions.

3.4 Endothelial Cells

Vascular endothelial cells are a rich source of t-PA in vivo and consequently have been extensively grown in vitro [69,70,71,72], as primary cultures, as models for studying t-PA synthesis and production. They have not been considered for large-scale production, despite the fact they specifically secrete t-PA, for several reasons. Firstly, until recently most of the investigations involving endothelial cells have been carried out with primary cells. However, improvements in isolation techniques [73] and growth medium formulations have now made available many cell lines isolated from aorta, vena cava, umbilical vessels and the retina. Secondly, endothelial cells produce a fast acting plasminogen activator inhibitor in excess of the free t-PA [74], (155 U ml^{-1}, 2.8 U ml^{-1} of active inhibitor after 24 h in culture) thus making measurement of t-PA impossible under certain conditions.

3.5 Recombinant Cells

The t-PA gene, isolated from the Bowes melanoma cell line, has been cloned into the plasmid pBR 322 and transformed into *Escherichia coli* [27]. However, the recombinant t-PA synthesised by *E. coli* [75] was not very active. The reasons for this are not clear but the absence of glycosylation and incorrect disulphide bonding may well be contributory factors. To overcome these problems the complete t-PA gene sequence has been engineered into mammalian cells. This was achieved by insertion of the gene into a vector, such as SV_{40}, and transfecting dihydrofolate reductase (DHFR) deficient Chinese hamster ovary (CHO) cells using the calcium phosphate precipitation method [51]. The mature protein plus signal presequence has 562 amino acids with approximately 251 amino acids of the serine protease portion plus a kringle sequence responsible for fibrin binding. A 10 [52] to 30 [51]-fold amplification has been achieved using methotrexate, an inhibitor of DHFR, resulting in yields of 50 pg d^{-1} cells [51]. Another approach has been to introduce additional copies of the t-PA gene into the Bowes melanoma cell line [48] and this recombinant cell line produced approximately 10-times more t-PA (3 pg per cell per day) and gave a t-PA harvest of 4000 IU ml^{-1} d^{-1}. Other examples of recombinant t-PA expressed in animal cells are the mouse L cell [53], and the CHO cell using adenovirus and DHFR amplification [50]. A comparison of yields of native and recombinant t-PA from these sources is given in Table 2.

Recombinant DNA technology and resulting amplification of t-PA expression now makes the economical production of therapeutic t-PA a reality but there is still scope for further increases in productivity by both genetic and phenotypic modifications. This is especially needed if clinical doses of 80 mg recombinant t-PA are required [54, 55].

3.6 Medium Formulation

t-PA producing cells are grown in conventional tissue culture media with various supplements. Serum is of importance for all the cell lines and needs to be included at 10% for maximum cell growth. During the enzyme production phase this concentration is reduced to 0.5–1% [46] or 0% [42,76] to facilitate enzyme purification from the supernatant. Melanoma cells can be kept in serum-free medium supplemented with insulin, transferrin, progesterone, cortisol and a mixture of trace elements for 6–8 weeks [76]. The medium was changed, and enzyme harvested, every 3 d but the cell loss that occurred each time necessitated a treatment with 5% foetal calf serum every 6–8 weeks to allow cell numbers to grow up again. As serum is so vital for growth of these plasminogen activator producing cells an important development is the method based on immuno affinity chromatography, which has been recently reported, that allows single-step purification from a supernatant containing 10% serum [77]. Production losses due to the "adherent" nature of the enzyme can be largely reduced by the inclusion of the surface active agent Tween 80 (0.01%) in both the culture medium and purification reagents. Loss of enzyme has been reported due to protease activity in the culture media and aprotinin (15 KIU ml^{-1}) is added by some investigators [62]. Many other media additions can be made in order to increase enzyme expression and these are reviewed in the next section.

3.7 Stimulation of t-PA Production

The yield of plasminogen activators from most cultured cells is extremely low and has always constituted a production problem, initially to get enough material for scientific investigation and currently to develop the enzyme as a therapeutic agent. Activator activity has been increased in culture by a wide range of compounds in a dose dependent manner and these are listed in Table 3. Many of these compounds are tumour-promoting agents and were studied because of the observation that cancer cells, or transformed

Table 3. Agents used for the stimulation of plasminogen activator expression from cell lines

Agent	Optimum concn.	Cell type	Stimulation	Ref.
5-Azacytidine	10–25 μM	BEB/GPK	×5	58)
Concanavalin A	50–100 $\mu g\ ml^{-1}$	Pig kidney	×5	107)
	20 $\mu g\ ml^{-1}$	HEL	×4	62)
	3×10^{-7} M	Macrophages	×10	14)
	50 $\mu g\ ml^{-1}$	GPK/BEB	×15	58)
Calcium	4.3 mM	Mouse 3T3	×6–30	36)
Calcitonin	0.1 $\mu g\ ml^{-1}$	LLC-PK	×5	83)
				49)
EGF	10 $ng\ ml^{-1}$	HE fibroblasts	×2–5	63)
	10 $ng\ ml^{-1}$	HeLa	×7	100)
Casein hydrolysate	1.5%	Hu fibroblasts	×7	64)
				71)
Fucosterol	25 μM	Bovine artery cells	×3–8	112)
cAMP	10^{-3} M	Rat granulosa	×3	82)
dbcAMP	10^{-3} M	Rat granulosa	×20	82)
Horse serum	10%	Hu. fibroblasts	×10	64)
Phytohaemagglutinin	30 $\mu g\ ml^{-1}$	Mouse macrophage	×2–3	14)
Poly-D-lysine	Surface coating	Human fibroblasts	×5	64)
Poly-D-lysine + horse serum	10%	Human fibroblasts	×20	64)
Phorbol myristate	30–100 $ng\ ml^{-1}$	CEF, RSVCEF	×8–12	113)
Acetate	1.5×10^{-8} M	CEF	×10	101)
	30–100 $mg\ ml^{-1}$	CEF	×10–30	65)
	100 $ng\ ml^{-1}$	HeLa	×10–20	101)
	10^{-9} M	Mouse macrophage	×2–10	14)
	50 nM	Endoth. vena cava	×3–20	72)
	10 $ng\ ml^{-1}$	HEF	×8–10	63)
	8×10^{-10} M	Granulocytes	×2–10	80)
	10 $ng\ ml^{-1}$	HeLa	×7	100)
	80–160 nM	Mouse muscle	×2–3	78)
	100 $ng\ ml^{-1}$	HEL	×5	114)
12-0-Tetradecanoyl	100 $ng\ ml^{-1}$	Bowes melanoma	×20	13)
Phorbol-13-acetate	5×10^{-8} M	Hamster HAK, DON	×11	115)
	100 $ng\ ml^{-1}$	GPK/BEB	×3	58)
	100 $ng\ ml^{-1}$	HeLa	×10–20	99)
Progesterone and 17 estradiol	1×10^{-8} M	Mammary human epithelial	×10	66)
Sitosterol	50 μM	Bovine artery	×3–8	112)
Retinoic acid	10^{-6} M	CEF	×9–10	106)
	10 μM	Mouse epith. tumour	×2–4	111)
	10 μM	Mouse melanoma	×2–4	111)
	10^{-6} M	HEF		106)
	10^{-6} M	Hu. oesophageal carcinoma	×15	106)
	10^{-6} M	Chick muscle	×2–5	78)
	10 μM	SV_{40} 3T3 mouse	×5–6	40)
Sodium butyrate	5–10 mM	Mouse tetratocarcinoma	×10	109)
		Mouse macrophages	×3	82)
Sulfonylureas	10 nM	Bovine aorta endothelial	×2–5	116)
Thrombin	2 $U\ ml^{-1}$	Human endothelial	×4	120)

counterparts of normal cells, had an approximately ten-fold higher secretion level of plasminogen activator. Examples of these agents are the phorbol esters, retinoids and mitogenic lectins. A number of phorbol esters have been investigated with varying effects on activator stimulation [70]. The most active are 12-0-tetradecanoyl phorbol-13-acetate (TPA) and phorbol myristate acetate. One analogue of TPA, phorbol didecanoate (PDD) was also found active but others (4-0-methyl-TPA and 4-PDD) were inactive [70,78]. The mode of action is believed to be at the transcriptional level by stimulating mRNA formation coding for t-PA, but its mitogenic effect could be due to activating the production of interleukin-2 [13]. Retinoic acid has been found more active than both retinol and retinyl acetate [65]. It has been observed in vivo that some steroid hormones enhance fibrinolysis [117] by increasing vascular t-PA activity [79]. One such hormone, stanozolol (stromba), has been found inactive in cultured cells but the list of agents in Table 3 include many hormones that have had a stimulatory effect. The list also includes agents which alter the culture environment to allow a higher presumed phenotypic expression of plasminogen activator, examples being poly-D-lysine and serum. Various other stimulants are effective because of their effect on the regulation of cell metabolism (e.g. the hypomethylating agent, azacytidine; epidermal growth factor (EGF); concanavalin-A). It is not usual for a mixture of compounds, even when of differing mode of action, to potentiate the response of a single agent [58,62]. A reason for this may be that, above a certain enzyme level, negative feedback controls operate [12].

Stimulation of plasminogen activator has been associated with changes in cAMP levels in cells. The actual significance of these changes is unknown because there are reports stating that cAMP levels should be low for plasminogen activator production whilst others indicate that raised levels are necessary. A consistent pattern is that macrophages [14,80] and fibroblast cells [81] require low cAMP levels and epithelial cells [82,83,84] high levels. There is a possible correlation to the fact that fibroblast and epithelial cells have different growth-regulatory responses to lectins and other compounds that effect secondary messenger (molecules within the cell modulated by surface changes which through a cascading chain of events control cell division e.g. cylic nucleotides) action [85,86], especially with regard to their adhesion properties. Low cAMP levels are a feature of malignant, transformed and dividing cells and can be induced by concanavalin A. Based on the fact that the two agents most effective in stimulating t-PA production, concanavalin A and phorbol ester, both reduce cAMP levels one has to conclude that low cAMP induces t-PA expression, at least in cells of non-malignant origin (e.g. fibroblasts and normal epithelial cells).

Stimulating agents have played a vital role in both allowing plasminogen activators to be produced in large enough quantities for characterisation studies and for elucidating regulatory processes involved in their production and mode of action. The degree of stimulation is not enough to allow an economical production process without going through the recombinant DNA process, but it is possible that recombinant cell lines may also be induced to produce higher levels of enzyme with some of these agents.

4 Purification of Tissue Plasminogen Activator

A survey of the literature over the past few years shows that t-PA has been isolated from many different organs and cells in culture. Table 4 lists purification methods for t-PA from some of these sources, together with their molecular weight, specific activity and the different methods employed to isolate the enzyme. This list is by no means complete as only the examples in which the enzymes has been purified to homogeneity, and characterised, are given.

The purification procedures used with various organs as the starting material, were lengthy and used very harsh conditions such as strong chaotropic solutions of potassium thiocyanate [6] or buffers of low pH [5]. These methods resulted in the purification of a few milligrams of enzyme from several kilogram quantity of starting material. With the discovery that malignant Bowes melanoma cells in culture secrete a plasminogen activator similar to human uterine t-PA, various cell lines have since been shown to secrete a single form of t-PA that is immunologically related to uterine and melanoma t-PA. Using the conditioned medium as the starting material, the purification methods have been relatively simple, utilising mainly an affinity chromatography and a gel filtration step [20,27,56,76,87,88]. With the availability of monoclonal antibodies to melanoma t-PA immunoaffinity chromatography has now been used as a single-step purification [77] method.

It was realised by many investigators that t-PA was proteolytically degraded during purification and resulted in a mixture of one and two chain forms. Hence a protease inhibitor, such as aprotinin [62], was used in all the buffers used for purification. Also, t-PA exhibits a strong adsorbing property to plastic and glass surfaces, resulting in significant losses during purification. This loss was minimised significantly by the use of low concentration of non-ionic detergents such as Tween 80.

The specific activities of t-PA listed in Table 4 are as quoted in the literature. They are expressed in different units and no attempt has been made to normalise them in the same units. However, comparison of t-PA activity, when expressed in the same units (against a reference International standard of human urokinase) also show wide variation with respect to different sources.

The molecular weight of purified plasminogen activators also varies considerably ranging from 35,000 to 85,000 Da. Such differences may be correlated to the diverse range of species, tissues and cell lines that have been used as a source of t-PA, but may also relate to limited proteolysis of the native protein.

5 Clinical Studies

Initially, the efficacy of t-PA as a specific fibrinolytic and thrombolytic agent was compared with human urokinase in an in vitro system composed of radiolabelled human blood clots suspended in circulating human plasma. t-PA derived from both melanoma [90] and epithelial [57] cell lines showed a dose-dependent lysis of blood clots and no significant fibrinogen breakdown in the surrounding plasma. It was also shown that non cross-linked clots lysed relatively more extensively than cross-linked clots. [91]

In view of these properties, t-PA was then evaluated for its in vivo thrombolytic activity in different animal models, with experimentally induced arterial or venous

Table 4. Purification of tissue plasminogen activator

Source	Specific activity	Molecular weight (Daltons)	Purification method	Ref.
Hog ovaries	100–175,000 units mg^{-1}	60,000	NH_4SCN extraction Acid ppt. Zinc fractionation Sephadex G-200 gel filtration	[6]
Human blood vessel perfusate	10–40,000 CTA units mg^{-1}	70–75,000	$(NH_4)SO_4$ fractionation Reverse $(NH_4)_2SO_4$ solubilization Octyl-Sepharose chrom. Sephadex G-75 gel filtration Sephadex G-150 gel filtration	[9]
Human uterus	16,000 IU mg^{-1}	64–69,000	Acetate buffer extraction $(NH_4)_2SO_4$ ppt. Zinc-chelate agarose chrom. n-Butyl agarose chrom. Con-A agarose chrom. Sephadex G-150 chrom.	[25]
Human blood vessel perfusate	50,000 CTA units mg^{-1}	56,000	Polyethylene glycol ppt. Hydroxyapatite chrom. Sepharose 6B gel filtration Lysine/fibrin agarose chrom.	[8]
Pig heart	250,000 IU mg^{-1}	64,000	$(NH_4)_2SO_4$ fractionation Adsorption on fibrin Sephacryl S-300 chrom. Arginine-Sepharose chrom. Sephacryl S-200 chrom.	[7]
Human melanoma cell	90,000 IU mg^{-1}	72,000	Zinc chelate agarose chrom. Con-A agarose chrom. Sephadex G-150 gel filtration	[20]
Human neuroblastoma cell	934,000 units mg^{-1}	37,500–66,500	$(NH_4)_2SO_4$ ppt. Affi-gel Blue chrom. p-Aminobenzaminidine Sepharose chrom.	[117]

Rat brain tumour cell	—	60,000	Zinc chelate agarose chrom. Con-A agarose chrom. Sepharose G-150 gel filtration	87)
Human embryonic lung (HEL)	10–40,000 CTA units mg^{-1}	56,000 & 83,000	Fibrin celite affinity chrom. Ultrogel ACA 44 gel filtration	62)
Human melanoma cell	220,000 IU mg^{-1}	72,000	Immunoaffinity chrom. Arginine-Sepharose chrom. Sephadex G-150 gel filtration	27)
RSV infected chick	477,000 units mg^{-1}	48,000	Fibrin celite affinity chrom. p-Aminobenzaminidine Sepharose chrom. Ultrogel ACA 22 gel filtration	89)
Human and guinea pig epithelial cell	6,000 & 12,500 IU mg^{-1}	62,000	Zinc chelate agarose chrom. Con-A agarose chrom. Ultrogel ACA 44 gel filtration	56)
Human melanoma cell	200,000 IU mg^{-1}	65,000	Immunosorbent chrom. Arginine Sepharose chrom. Sephadex G-150 gel filtration	88)
Human melanoma cell	80–100,000 IU mg^{-1}	67,000	SP Sephadex chrom. Sephadex G-100 gel filtration	76)

thrombi. A detailed study in rabbit with a thrombus in the jugular vein showed that the thrombolysis achieved with t-PA was specific and dose-dependent and was devoid of any extensive systemic fibrinolytic activation [92]. A similar study with epithelial t-PA basically reached the same conclusion [60]. The study also showed that local infusion of t-PA in the vicinity of the thrombus was relatively more effective than systemic infusion, but the degree of thrombolysis was much less affected by the age of the thrombus.

In another study, lysis of coronary thrombi with t-PA was investigated in dogs [93] and baboons [94]. An arterial thrombus was formed with a copper coil in the left anterior descending coronary artery. It was shown that intravenous infusion with melanoma t-PA resulted in timely reperfusion of the occluded vessel with substantial salvage of the myocardial tissue and restoration of intermediary metabolism.

The first successful therapeutic use of t-PA as a thrombolytic agent was carried out in 1981 when two patients with renal and ileofemoral thrombosis, confirmed by angiography, were treated intravenously with 7.5 mg of purified enzyme over 24 h [47]. The occluded vessel was reperfused without any haemorrhagic disorder or haemostatic breakdown. This report confirmed the results obtained in animal experiments and thus opened a new approach to the treatment of various thromboembolic diseases.

With the availability of significant amounts of purified t-PA from large-scale tissue culture methods, and with increasing evidence obtained in animal studies for its clot-selectivity and lack of side effects, attempts were made to lyse coronary thrombi with t-PA in seven patients with evolving myocardial infarction [95]. Infusion of t-PA (0.3 to 1.4×10^6 IU), by intracoronary or intravenous routes, induced thrombolysis within 19 to 50 min in six of the seven patients. The reperfusion of the occluded vessel was confirmed by angiography in each case. Investigation of the haemostatic parameters also confirmed that there was no serious systemic fibrinolytic activation.

Another noteable achievement in the success of t-PA has been the cloning and expression of the t-PA gene in a mammalian cell system and its availability in sufficient amounts for clinical studies [75]. This led to two randomised, multicentre clinical trials to assess and compare the efficacy of intravenous t-PA and streptokinase in patients with acute myocardial infarction.

One of these trials, the Thrombolysis In Myocardial Infarction (TIMI) Trial [55], carried out entirely in the United States, recently published its Phase I findings showing substantial, statistically significant differences in recanalisation rates between the patients given t-PA and those given streptokinase. A similar conclusion was also reached by the European Cooperative Study Group [96] in their single-blind randomised trial of intravenous recombinant t-PA versus intravenous streptokinase in acute myocardial infarction. Phase II of the TIMI trial is a randomised placebo-controlled trial which is presently ongoing. Detailed summaries of these, and other trials, have been reviewed by Boissel [54].

6 Conclusion

This review on the study of plasminogen activator from the early work in small scale tissue culture through to a recombinant DNA product being prepared for clinical trial is a good reflection of the progression and impact of Biotechnology in science today. Initially improvements in cell culture procedures and the derivation of new cell lines allowed increased production of material so that biochemical and limited clinical characterisation studies could be carried out. This allowed the recognition and definition of the tissue and urokinase type activators and led to the need for larger quantities of t-PA for clinical evaluation. The classical problem of the very low productivity of animal cell culture systems had then to be overcome. Initially the easy growth and elevated levels of t-PA expression of the Bowes melanoma cell line made enough material available for limited clinical investigations. The role of fibrinolysis in cancer is a subject of intensive study and useful information has been obtained for amplifying cell yields of t-PA using tumour promoting agents and various growth regulating compounds. By this means the use of normal (i.e. non-tumourgenic) cell lines became more feasible as the characteristic lower yields of these cells could be increased to levels similar to the malignant Bowes melanoma cells.

It was very clear that if t-PA was to be used as a therapeutic agent then a production process would have to be based on recombinant DNA technology. This approach was needed both to make large scale production economically possible and to avoid the use of a malignant cell substrate for a clinical product. Genentech Inc. pioneered the way and patented the use of the Bowes melanoma t-PA gene expression in both bacterial *(E. coli)* and mammalian cells (CHO). This route, or similar routes have now been followed by many other commercial companies. An alternative approach has been to use a structurally modified t-PA by degrading or removing part of the carbohydrate portion [97)], or by forming an acyl-streptokinase plasminogen complex [98)]. The main aim of both these products was to produce a molecule with an increased biological half-life.

With over 20 major pharmaceutial companies worldwide working on a clinical preparation of t-PA, and the encouraging results from limited clinical trials, the prospect over the next 5–10 years for reducing thrombolytic disorders is very great. Its widespread use will depend upon achieving safe and economical production and as yet there is not enough data to calculate the optimum dose size and regime. There is strengthening evidence that there may be a family of t-PA's as indicated by the differences in melanoma and epithelial cell t-PA (the only sources besides human to be fully characterised), which will have specific clinical targets.

In conclusion, modern biotechnological procedures have enabled a research chemical to become a therapeutic agent, and as scale-up technologies improve, this product will have a huge impact on worldwide health care over the next decade.

7 References

1. Collen, D.: Thromb. Haemost. *43*, 77 (1980)
2. Rijken, D. C., Wijngaards, G., Welbergen, J.: Thromb. Res. *18*, 815 (1980)
3. Rijken, D. C., Wijngaards, G., Welbergen, J.: J. Lab. Clin. Med. *97*, 477 (1981)
4. Astrup, T.: In Proteases and Biological Control (eds. Reich, E., Rifkin, D. B.) Vol. 2, 343. Cold Spring Harbour Laboratory, N.Y. 1975
5. Rijken, D. C., Wijngaards, G., Zaak de Jong, M., Welbergen, J.: Biochim. Biophys. Acta. *580*, 140 (1979)
6. Kok, P., Astrup, T.: Biochem. *8*, 79 (1979)

7. Wallen, P., Beigsdorf, N., Ranby, M.: Biochim. Biophys. Acta. *719*, 318 (1982)
8. Allen, R. A., Pepper, D. S.: Thromb. Haemost. *45*, 43 (1981)
9. Binder, B. R., Spragg, J., Austen, K. F.: J. Biol. Chem. *254*, 1998 (1979)
10. Todd, A. S.: J. Path. Bact. *78*, 281 (1959)
11. Emeis, J.: Thromb. Res. *30*, 195 (1983)
12. Kadouri, A., Bohak, Z.: Adv. Biotech. Proc. *5*, 275 (1985)
13. Opdendakker, G., Ashino-Fuse, H., Van Damme, J., Billiau, A., De Sommer, P.: Eur. J. Biochem. *131*, 481 (1983)
14. Vassali, J. D., Hamilton, J., Reich, E.: Cell *11*, 695 (1977)
15. Werb, Z., Mainardi, C. L., Vater, C. A., Harris, E. D.: New Eng. J. Med. *296*, 1017 (1977)
16. Bosmann, H. B., Hall, T. C.: Proc. Natl. Acad. Sci. *71*, 1833 (1974)
17. Boyd, J. B., Farb, R. H., Yost, F. J., Georgiade, N., Lazarus, G. S.: Surg. Forum *28*, 131 (1977)
18. Beers, W. H.: Cell *6*, 379 (1975)
19. Ossowski, L., Biegel, D., Reich, E.: ibid. *16*, 929 (1979)
20. Rijken, D. C., Collen, D.: J. Biol. Chem. *256*, 7035 (1981)
21. Banyai, L., Varadi, A., Patthy, L.: FEBS Letts. *163*, 37 (1983)
22. Strassburger, W., Wollmer, A., Pitt, J. E.: ibid. *157*, 219 (1983)
23. Ranby, M., Bergsdorf, N., Nilsson, T.: Thromb. Res. *27*, 175 (1982)
24. Rijken, D. C., Hoylaerts, M., Collen, D.: J. Biol. Chem. *257*, 2920 (1982)
25. Ranby, M., Bergsdorf, N., Pohl, G., Wallen, P.: FEBS Letts. *146*, 289 (1982)
26. Bennett, W. F.: Thromb. Haemost. *50*, 106 (1983)
27. Wallen, P., Pohl, G., Bergsdorf, N., Ranby, M., Ny, T., Jornvall, H.: Eur. J. Biochem. *132*, 681 (1983)
28. Bernik, M. B., Kwaan, H. C.: J. Clin. Invest. *48*, 1740 (1969)
29. Leuvis, L. J.: Thromb. Haemost. *42*, 895 (1979)
30. Lewis, M. L., Barlow, C. H., Morrison, D. R., Nacchtwey, D. S., Fessler, D. L.: Haemostasis *11*, 43 (1982)
31. Wilson, E. L., Becker, M. L. B., Hoal, E. G., Dowdle, E. B.: Cancer Res. *40*, 933 (1980)
32. Wilson, E. L., Dowdle, E.: Int. J. Cancer *22*, 390 (1978)
33. Levin, E. G., Loskutoff, D. J.: J. Cell Biol. *94*, 631 (1982)
34. Kirchleimer, J., Koller, A., Binder, B. R.: Biochim. Biophys. Acta. *797*, 256 (1984)
35. Wilson, E. L., Jacobs, P., Dowdle, E. B.: Blood *61*, 568 (1983)
36. Chou, I. N., Roblin, R. O., Black, P. H.: J. Biol. Chem. *252*, 6256 (1977)
37. Ossowski, L., Unkless, J. C., Tobia, A., Quigley, D. B., Reich, E. J.: J. Exp. Med. *137*, 112 (1973)
38. Rifkin, D. B., Pollack, R.: J. Cell Biol. *73*, 47 (1975)
39. Rohrlich, S. T., Rifkin, D. B.: ibid. *75*, 31 (1977)
40. Schroder, E. W., Chou, I. N., Black, P. H.: Cancer Res. *40*, 3089 (1980)
41. Robbin, R. D., Young, P. L., Bell, T. E.: Biochem. Biophys. Res. Comm. *82*, 165 (1978)
42. Griffiths, J. B., McEntee, I. D., Electricwala, A., Atkinson, A., Sutton, P. M., Naish, S., Riley, P. A.: Develop. Biol. Standard *60*, 439 (1985)
43. Astrup, T.: In Thrombosis and Urokinase (eds. Paoletti, R., Sherry, S.) Vol. 11, 1979
44. Vetterlein, D., Young, P. L., Bell, T. E., Roblin, R.: J. Biol. Chem. *254*, 575 (1979)
45. Collen, D., Rijken, D. C., Van Damme, J., Billiau, A.: Thromb. Haemost. *48*, 294 (1982)
46. Kluft, C., van Wezwl, A. L., van der Velden, C. A. N., Emeis, J. J., Verheijen, J. H., Wijnmgaards, G.: Adv. in Biotech. Proc. *2*, 508 (1983)
47. Weimer, W., Stibbe, J., Van Seyen, A. J., Billiau, A., De Somer, P., Collen, D.: Lancet *2*, 1018 (1981)
48. Browne, M. J., Dodd, I., Carey, J. E., Chapman, C. G., Robinson, J. H.: Thromb. Haemost. *54*, 422 (1985)
49. Dayer, J. M., Vassalli, J. D., Bobbitt, L., Hull, R. N., Reich, E., Krane, S. M.: J. Cell Biol. *91*, 195 (1981)
50. Kaufman, R. J., Wasley, L. C., Spiliotes, A. J., Gossels, S. D., Latt, S. A., Larsen, G. A., Kay, R. M.: Mol. Cell Biol. *5*, 1750 (1983)
51. Geoddel, D. V. N., Kohr, W. J., Pennica, D., Vehar, G. A.: U.K. Patent Appl. GB2119804A (1983)
52. Zamarron, C., Lijnen, H. R., Collen, D.: J. Biol. Chem. *259*, 2080 (1984)
53. Browne, M. J., Tyrrell, A. W. R., Chapman, C. G., Carey, J. E., Glover, D. M., Grosweld, F. G., Dodd, I., Robinson, J. H.: Gene *33*, 279 (1985)

54. Boissel, J. P.: Thromb. Haemost. *55*, 282 (1986)
55. Thrombolysis in Myocardial Infarction (TIMI) trial. New Eng. J. Med. *312*, 932 (1985)
56. Electricwala, A., Atkinson, T.: Eur. J. Biochem. *147*, 511 (1985)
57. Electricwala, A., Ling, R. J., Sutton, P. M., Griffiths, B., Riley, R. A., Atkinson, T.: Thromb. Haemost. *53*, 200 (1985)
58. Electricwala, A., Griffiths, B.: Cell Biochem. Function *4*, 55 (1986)
59. Griffiths, J. B., Atkinson, T., Electricwala, A., Latter, A., Ling, R., McEntee, I. D., Riley, P. A., Sutton, P. M.: Develop. Biol. Standard *55*, 31 (1984)
60. Electricwala, A., Emeis, J. J., Atkinson, T.: J. Pharm. Exp. Therap. *238*, 254 (1986)
61. Atkinson, A., Electricwala, A., Latter, A., Riley, P. A., Sutton, P. M.: Lancet, vol. *2*, 132 (1982)
62. Brounty-Boyle, G. C., Maman, M., Marian, J. C., Choay, P.: Biotechnology *2*, 1058 (1984)
63. Eaton, D. L., Baker, J. B.: J. Cell Biol. *97*, 323 (1983)
64. Kadouri, A., Bohak, Z.: Biotechnology *1*, 354 (1983)
65. Wilson, E. L., Dowdle, E. B.: Cancer Res. *40*, 4817 (1980)
66. Yang, N. S., Kirland, W., Jorgenson, T., Furmauski, P. J.: J. Cell Biol. *84*, 120 (1980)
67. Crutchley, D. J., Conanan, L. B., Maynard, J. R.: Fed. Proc. *42*, 591 (1983)
68. Hamilton, J. A.: Arthritis. Rheum. *25*, 432 (1982)
69. Thorsen, S., Philys, M.: Haemostasis *11*, 4 (1982)
70. Gross, G. L., Moscattelli, D., Jaffee, E. A., Rijken, D. B.: J. Cell Biol. *95*, 974 (1982)
71. Shimonaka, M., Hagiwara, H., Kojima, S., Inada, T.: Thromb. Res. *36*, 217 (1984)
72. Loskutoff, D. J., Edginglon, T. S.: Proc. Natl. Acad. Sci. *74*, 3903 (1977)
73. Ziats, N. P., Gioldsmith, G. H., Medvedeff, E. D., Robertson, A. L.: Thromb. Res. *41*, 239 (1986)
74. Levin, E. G.: Blood *67*, 1309 (1986)
75. Pennica, D., Holmes, W. E., Kohr, W. J., Harkins, R. N., Vehar, G. A., Ward, C. A., Bennett, W. F., Yelverton, E., Seeburg, P. H., Heyneker, H. L., Geoddel, D. V., Collen, D.: Nature, *310*, 214 (1983)
76. Kruithof, E. K. O., Schlwening, W. D., Bachmann, F.: Biochem. J. *226*, 631 (1985)
77. Reagan, M. E., Robb, M., Bornstein, I., Niday, E. G.: Thromb. Res. *40*, 1 (1985)
78. Miskin, R., Easton, T. G., Reich, E.: Cell *15*, 1301 (1978)
79. Preston, F. E.: Scottish Med. J. *59*, (1981)
80. Granelli-Piperno, A., Vassali, J. D., Reich, E.: J. Exp. Med. *146*, 1693 (1977)
81. Wilson, E. L., Reich, E.: Cancer Res. *39*, 1579 (1979)
82. Beers, W. H., Strickland, S., Reich, E.: Cell *6*, 387 (1976)
83. Nagamine, Y., Sudal, M., Reich, E.: ibid. *32*, 1181 (1983)
84. Lacroix, M., Smith, F. E., Fritz, I. B.: Mol. Cell Endocrinol. *9*, 227, (1977)
85. Grinelli, F.: Exp. Cell Res. *97*, 265 (1976)
86. Prinz, R., van Figura, K.: ibid. *112*, 275 (1978)
87. Bykowska, K., Rijken, D. C., Collen, D.: Thromb. Haemost. *46*, 642, (1981)
88. Einarsson, M., Brandt, J., Kaplan, L.: Biochim. Biophys. Acta. *1*, 830 (1985)
89. Kacian, D. L., Harvey, R. C.: Arch. Biochem. Biophys. *236*, 354 (1985)
90. Matsuo, O., Rijken, D. C., Collen, D.: Thromb. Haemost. *45*, 225 (1981)
91. Korningen, C., Collen, D.: ibid. *46*, 561 (1981)
92. Collen, D., Strassen, J. M., Verstraete, M.: J. Clin. Invest. *71*, 368, (1983)
93. Bergman, S. R., Fox, K. A. A., Ter-Pogossian, M. M., Sobel, B. E., Collen, D.: Science *220*, 1181 (1983)
94. Flameng, W., Van der Werf, F., Vanhacke, J., Verstraete, M., Collen, D.: J. Clin. Invest. *75*, 84 (1985)
95. Van der Werf, F., Ludbrook, P. A., Bergmann, S. R.: New Eng. J. Med. *310*, 609 (1984)
96. Verstraete, M.: Lancet 842 (1985)
97. Robinson, J. H.: Patent Appl. No. PCT/GB83/00273 (1983)
98. Smith, R. A. G., Dupe, R. J., English, B. D., Green, J.: Nature *290*, 505 (1981)
99. Schleuning, W. D.: Thromb. Haemost. *50*, 56 (1983)
100. Lee, L. S., Veinstein, I. B.: Nature *274*, 696 (1978)
101. Wigler, M., Weinstein, I. B.: ibid. *259*, 232 (1976)
102. Kjaldgaard, A., Lansson, B., Alstedt, B.: Thromb. Res. *36*, 591 (1984)
103. Gronow, M., Bliem, R.: Trends in Biotech. *1*, 26 (1983)

104. Jones, P. A., Laug, W. E., Benedict, W. F.: Cell *6*, 245 (1975)
105. Granelli-Piperno, A., Reich, E.: J. Exp. Med. *148*, 223 (1978)
106. Wilson, E. L., Reich, E.: Cell *15*, 385 (1978)
107. Mochan, E.: Biochim. Biophys. Acta. *588*, 273 (1979)
108. Strickland, W., Mahdrau, M.: Cell *15*, 393 (1978)
109. Ichikawa, N., Miyashita, T., Nishimune, Y., Matsushiro, A., Biken, J.: *27*, 143 (1984)
110. Mak, T. W., Rutledge, G., Sutherland, D. J. A.: Cell *7*, 223 (1976)
111. Lotan, R., Lotan, D., Kadouri, A.: Exp. Cell Res. *141*, 79 (1982)
112. Hagiwara, H., Shimonaka, M., Marisaki, M., Ikekawa, N., Inada, Y.: Thromb. Res. *33*, 363 (1984)
113. Goldfarb, R. H., Quigley, J. P.: Cancer Res. *38*, 4601 (1978)
114. Jaken, S., Geffen, C., Black, P. H.: Biochim. Biophys. Acta. *99* 379 (1981)
115. Christman, J. K., Copp, R. P., Pedrinan, L., Whalen, C. E.: Cancer Res. *38*, 3854 (1978)
116. Kuo, B. S., Korner, G., Bjornsson, T. D.: In Cardiovascular Disease '86 Abst. 222 (1986)
117. Noll, G., Lammle, B., Duckert, F.: Thromb. Res. *37*, 529 (1985)
118. Vassali, J. D., Hamilton, J., Reich, E.: Cell *8*, 271 (1976)
119. Electricwala, A., Sutton, P. M., Griffiths, B., Ríley, P. A., Atkinson, T.: Prog. in Fibrinolysis *7*, 341 (1985)
120. Gelehrter, T. D., Sznycer-Laszuk, R. J.: J. Clin. Invest. *77*, 165 (1986)

MIX

Papier | Fördert gute Waldnutzung

FSC® C083411

Zeitfracht Medien GmbH
Ferdinand-Jühlke-Straße 7
99095 Erfurt, Deutschland
produktsicherheit@kolibri360.de